T0144579

INTRODUCTION TO
BIOLOGICAL NETWORKS

CHAPMAN & HALL/CRC
Mathematical and Computational Biology Series

Aims and scope:
This series aims to capture new developments and summarize what is known over the entire spectrum of mathematical and computational biology and medicine. It seeks to encourage the integration of mathematical, statistical, and computational methods into biology by publishing a broad range of textbooks, reference works, and handbooks. The titles included in the series are meant to appeal to students, researchers, and professionals in the mathematical, statistical and computational sciences, fundamental biology and bioengineering, as well as interdisciplinary researchers involved in the field. The inclusion of concrete examples and applications, and programming techniques and examples, is highly encouraged.

Series Editors

N. F. Britton
Department of Mathematical Sciences
University of Bath

Xihong Lin
Department of Biostatistics
Harvard University

Hershel M. Safer
School of Computer Science
Tel Aviv University

Maria Victoria Schneider
European Bioinformatics Institute

Mona Singh
Department of Computer Science
Princeton University

Anna Tramontano
Department of Biochemical Sciences
University of Rome La Sapienza

Proposals for the series should be submitted to one of the series editors above or directly to:
CRC Press, Taylor & Francis Group
3 Park Square, Milton Park
Abingdon, Oxfordshire OX14 4RN
UK

Published Titles

Published Titles (continued)

Kinetic Modelling in Systems Biology
Oleg Demin and Igor Goryanin

Knowledge Discovery in Proteomics
Igor Jurisica and Dennis Wigle

Meta-analysis and Combining Information in Genetics and Genomics
Rudy Guerra and Darlene R. Goldstein

Methods in Medical Informatics: Fundamentals of Healthcare Programming in Perl, Python, and Ruby
Jules J. Berman

Modeling and Simulation of Capsules and Biological Cells
C. Pozrikidis

Niche Modeling: Predictions from Statistical Distributions
David Stockwell

Normal Mode Analysis: Theory and Applications to Biological and Chemical Systems
Qiang Cui and Ivet Bahar

Optimal Control Applied to Biological Models
Suzanne Lenhart and John T. Workman

Pattern Discovery in Bioinformatics: Theory & Algorithms
Laxmi Parida

Python for Bioinformatics
Sebastian Bassi

Quantitative Biology: From Molecular to Cellular Systems
Sebastian Bassi

Spatial Ecology
Stephen Cantrell, Chris Cosner, and Shigui Ruan

Spatiotemporal Patterns in Ecology and Epidemiology: Theory, Models, and Simulation
Horst Malchow, Sergei V. Petrovskii, and Ezio Venturino

Statistics and Data Analysis for Microarrays Using R and Bioconductor, Second Edition
Sorin Drăghici

Stochastic Modelling for Systems Biology, Second Edition
Darren J. Wilkinson

Structural Bioinformatics: An Algorithmic Approach
Forbes J. Burkowski

The Ten Most Wanted Solutions in Protein Bioinformatics
Anna Tramontano

Chapman & Hall/CRC Mathematical and Computational Biology Series

INTRODUCTION TO BIOLOGICAL NETWORKS

ALPAN RAVAL

ANIMESH RAY

CRC Press
Taylor & Francis Group
Boca Raton London New York

CRC Press is an imprint of the
Taylor & Francis Group, an **informa** business

A CHAPMAN & HALL BOOK

CRC Press
Taylor & Francis Group
6000 Broken Sound Parkway NW, Suite 300
Boca Raton, FL 33487-2742

First issued in paperback 2019

© 2013 by Taylor & Francis Group, LLC
CRC Press is an imprint of Taylor & Francis Group, an Informa business

No claim to original U.S. Government works

ISBN-13: 978-1-58488-463-7 (hbk)
ISBN-13: 978-0-367-38013-7 (pbk)

This book contains information obtained from authentic and highly regarded sources. Reasonable efforts have been made to publish reliable data and information, but the author and publisher cannot assume responsibility for the validity of all materials or the consequences of their use. The authors and publishers have attempted to trace the copyright holders of all material reproduced in this publication and apologize to copyright holders if permission to publish in this form has not been obtained. If any copyright material has not been acknowledged please write and let us know so we may rectify in any future reprint.

Except as permitted under U.S. Copyright Law, no part of this book may be reprinted, reproduced, transmitted, or utilized in any form by any electronic, mechanical, or other means, now known or hereafter invented, including photocopying, microfilming, and recording, or in any information storage or retrieval system, without written permission from the publishers.

For permission to photocopy or use material electronically from this work, please access www.copyright.com (http://www.copyright.com/) or contact the Copyright Clearance Center, Inc. (CCC), 222 Rosewood Drive, Danvers, MA 01923, 978-750-8400. CCC is a not-for-profit organization that provides licenses and registration for a variety of users. For organizations that have been granted a photocopy license by the CCC, a separate system of payment has been arranged.

Trademark Notice: Product or corporate names may be trademarks or registered trademarks, and are used only for identification and explanation without intent to infringe.

Library of Congress Cataloging-in-Publication Data

Ray, Animesh, 1954-
    Introduction to biological networks / Animesh Ray, Alpan Raval.
       pages cm. -- (Chapman & Hall/CRC mathematical & computational biology)
    Includes bibliographical references and index.
    ISBN 978-1-58488-463-7 (hardback)
    1. Biological systems--Mathematical models. 2. Systems biology--Mathematical models. 3. Computational biology. I. Raval, Alpan, 1968- II. Title.

QH323.5.R38 2013
571.7--dc23   2013003654

Visit the Taylor & Francis Web site at
http://www.taylorandfrancis.com

and the CRC Press Web site at
http://www.crcpress.com

Contents

Preface

In the 1940s and 1950s, biology was transformed by physicists and physical chemists, who employed simple yet powerful concepts and engaged the powers of genetics to infer mechanisms of biological processes. The biological sciences borrowed from the physical sciences the notion of building intuitive, testable, and physically realistic models by reducing the complexity of biological systems to the components essential for studying the problem at hand. Molecular biology was born.

A similar migration of physical scientists and of methods of physical sciences into biology has been occurring in the decade following the complete sequencing of the human genome, whose discrete character and similarity to natural language have additionally facilitated the application of the techniques of modern computer science. Furthermore, the vast amount of genomic data spawned by the sequencing projects has led to the development and application of statistical methods for making sense of this data. The sheer amount of data at the genome scale that is available to us today begs for descriptions that go beyond simple models of the function of a single gene to embrace a system-level understanding of large sets of genes functioning in unison. It is no longer sufficient to understand how a single gene mutation causes a change in its product's biochemical function, although this is in many cases still an important problem. It is now possible to address how the consequences of a mutation might reverberate through the interconnected system of genes and their products within the cell.

The scale of a system-level description of genes necessitates an approximate treatment of the biochemical entities involved: the DNA and RNA molecules, and the enzymes that interact with them to alter their levels of activity. In the simplest approximation, these entities can be described as featureless *nodes* in a genome-scale network whose *edges* represent various types of interaction between the nodes. This network-based description appears to dramatically reduce the complexity of biology to the formalism of idealized physical systems, computing systems, their architecture, and their dynamical properties.

Today the modeling of biological processes can typically involve the simultaneous description of tens of thousands of genes and their products; the set of interactions among them has become an equally important object of study as the set of biological entities themselves. This new way of thinking about biological complexity requires new theoretical and quantitative concepts and methodologies, and biology is once again attracting students and practitioners from the mathematical, physical, and computer sciences. Concurrently, biology curricula in many educational institutions are being augmented by mathematics and computer science courses. It is hoped that this book will serve as an entry point for both these groups of students to the key concepts in modern, genomics-inspired network biology.

This book is about the networks of interaction among biological molecules—the experimental methods of uncovering and testing these networks (Chapters 1, 2, 5, and 6), computational methods for predicting them (Chapters 3 and 4), general mechanisms of network formation and evolution (Chapter 7), and the application of network approaches to important problems in biology and medicine (Chapters 8 and 9). It should be accessible to students with undergraduate degrees in the physical, mathematical, and computer sciences with some exposure to molecular biology, as well as to students with undergraduate degrees in biology with exposure to calculus and probability and linear algebra. There are portions of the book that will challenge both sets of readers: for example, Chapter 6 is expected to be difficult for the mathematically inclined student with little exposure to biology, while the biology student might find Chapter 7 daunting.

Network biology is a large field that is very much in a state of flux as new directions emerge. While we have tried to acquaint our readers with many areas of modern research, the choice of topics covered necessarily reflects our bias and expertise. There are important areas that we do not cover which we think are very well addressed in other textbooks. These include detailed treatments of gene regulatory dynamics and the role of network motifs, networks of cells including networks of nerve cells, genome-wide disease association studies, and disease models. Overall, the book emphasizes concepts over techniques; the reader will not find detailed experimental protocols or computer algorithms here but it is our hope that she will come away with a conceptual understanding of both.

This book would not have adopted its present form without the constructive criticism and help offered by students and colleagues. We would especially like to thank Sonali Talele, Mitsunori Ogihara, and Carsten Peterson for comments on a preliminary draft of the book. Sonali Talele contributed Figure 9.3, and helped in seeking permissions for the reproduction of figures. Jesse Frumkin contributed to Section 6.4. Sonal Majeethia helped with other figures, and Roma Panjwani with references. Roshni Ray-Ricchetti provided feedback on an early version of Chapter 6. Krishanu Ray served as a sounding board for many of the ideas during long drives with one of us. Rachel Holt kept us organized. Perhaps like many first-time authors, we vastly underestimated the time and effort needed to undertake this ambitious project. This has been all the more challenging because the topics of this book are inherently interdisciplinary. The book would not have been completed were it not for the constant encouragement and support of our editor, Sunil Nair. Needless to say, any errors that survived our endless revisions are the sole responsibility of the authors.

Last but not least, we thank our spouses Sonal and Sumita for deep understanding and sacrifice, without which this book would not be.

Alpan Raval
Animesh Ray

Chapter 1

The Living Interactome

Biology presents many perplexing themes to one trained in the physical or mathematical sciences. It has very few general laws or universal rules. While evolutionary theory represents a general framework, the nature of evolutionary forces and the precise entities upon which these forces act are matters of debate. The genetic code, universal for terrestrial life, appears to be a historical accident in the sense that among several plausible alternative codes, only the present coding scheme has survived through evolution. Almost everything else in biology depends on the context; every rule has its exception.[1]

One of the most puzzling aspects of biology is that the line between a living cell and its "non-living" parts is fuzzy at best. Is a mitochondrion alive? This intracellular organelle is ubiquitous in cells that possess nuclei and live in an oxygen-rich environment. Mitochondria apparently evolved through a symbiosis of ancient bacteria-like organisms (called *Eubacteria* or *Proteobacteria*) with another class of organisms called the *Archaea* (Box 1.1 and Figure 1.1). In this process, interactions and communication among genes and proteins needed to be reestablished in the two organisms so as to make the symbiosis possible. The symbiosis presumably involved losing certain genes that are required for independent existence and gaining some that are needed for a mutually dependent and entangled way of life for the two partners. Moreover, specialized functions of certain genes needed to be generalized, and new specialized genes needed to have evolved.

If mitochondria are not themselves alive, then at what stage in the process of establishing a mutually dependent relationship with another organism did a *prokaryote* (Box 1.1) lose its "living" status? More generally, it is unclear how much of the biological world is a result of historical accident and how much follows inexorable deterministic

---

[1] A similar view was best expressed by Max Delbrück in an essay entitled "A Physicist Looks at Biology" (Delbrück, 1966).

Box 1.1 Eukaryotes, Prokaryotes, and Archaea.

Eukaryotes are organisms that have a nucleus in each of their constituent cells. The nucleus contains the chromosomal DNA. Eukaryotes can be unicellular or multicellular. Most "higher" organisms, such as flowering plants, mammals, etc., are eukaryotes, as are certain unicellular plants (some algae), animals (e.g., the malaria-causing parasite *Plasmodium falciparum*), and fungi (e.g., the baker's yeast *Saccharomyces cerevisiae*). Prokaryotes are organisms that do not possess a nucleus within their cells. Their genomic DNA is not compartmentalized. Prokaryotes are almost always unicellular, and consist of two major types: the eubacteria and the archaebacteria (or archaea). Eubacteria are found almost everywhere on Earth, whereas archaebacteria usually occur in extreme environments, such as thermal hot springs and highly saline lakes. Archaebacterial DNA is compositionally different from that of eubacterial DNA. The chromosomes of eubacteria are circular, whereas the linear archaebacterial chromosomes resemble the linear chromosomes of eukaryotes. Eukaryotes share many features of both archaebacteria (such as RNA synthesis mechanisms and chromosome structure) and eubacteria (such as DNA replication and metabolic pathways).

---

laws. Nonetheless, progress over the past fifty years in understanding the fundamental basis of life has profited from one general idea: the set of genes (*genotype*) in an individual interacts with the environment in a well-defined way to determine the *phenotype* (the set of observable properties of the organism), and that evolution acts on the phenotype. We now understand that the phenotype is the result of numerous physical interactions among molecules comprising the cell, which are mostly controlled or encoded for by genes. It is therefore natural that the most successful approach to understanding biology has been through the reductionist's method of describing the parts list of the cell (see Box 1.2 for a summary of the very basic molecular biological concepts). These parts are the genes, the proteins, and the metabolites. Insights into biological processes must ultimately come from mechanistic models involving these molecules and their interactions. This is the view of a molecular biologist.

It is becoming clear, however, that a detailed mechanistic description of each interaction sometimes clouds understanding of the biological relevance of that interaction. What is often more important is to place that interaction within the context of other interactions. For example, consider the case of transmission of a pheromone signal (Box 1.3 and

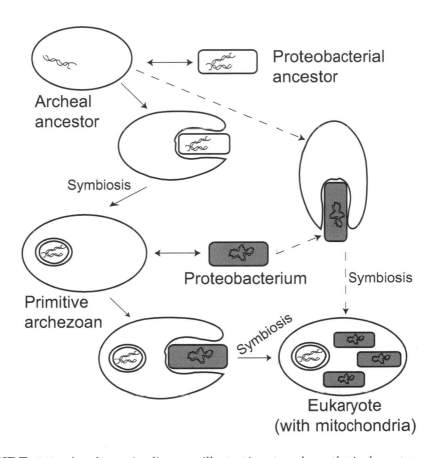

FIGURE 1.1: A schematic diagram illustrating two hypothetical routes for the evolution of eukaryotic cells with mitochondria. The first model proposes that an ancestral archaebacterium engulfed an ancestral eubacterium (proteobacterium); the genetic materials of the two contributed to the formation of the first mitochondria-less eukaryote (a primitive archezoan), which later fused with another eubacterium (labeled in the figure as an α-protcobacterium), thus giving rise to the mitochondriated eukaryotic cell. The second model proposes that the evolution of the eukaryotic cell occurred in one step by the fusion of a hydrogen-requiring archaebacterium and a hydrogen-producing α-proteobacterium. These symbiotic relationships presumably required the evolution of numerous novel interactions between genes, proteins, and other molecules produced by different organisms. Adapted with permission from Gray et al. (1999).

Box 1.2 A few concepts in molecular biology.

The central dogma of molecular biology states that information encoded in the DNA sequence determines the RNA sequence (this process is called *transcription*), which in turn determines the amino acid sequence of proteins (*translation*) due to the correspondence between each triplet of RNA nucleotides and a single amino acid. This correspondence is the *genetic code*. In addition, however, information can be transferred from DNA to DNA and RNA to RNA by *replication*, and from RNA to DNA by *reverse transcription*. No mechanism of information transfer from protein to DNA or RNA is known. However, an RNA or protein molecule can modulate the amount of RNA or protein transcribed or translated from another gene. This amount is called the *expression level* of the corresponding RNA or protein, and the RNA or protein molecule that modulates the expression level is called a *regulator*. Regulators can be positive or negative, depending on whether the expression level of their target is increased or decreased, respectively. Proteins that regulate the expression of a gene are called *transcription factor* proteins.

RNA is transcribed from *gene* sequences. For example, RNA that is eventually translated to protein is called messenger RNA, or simply mRNA. The RNA from certain genes is never translated into protein. Examples of such genes are those encoding ribosomal RNA (rRNA), transfer RNA (tRNA), microRNA (miRNA), etc. The set of expression levels of RNA transcribed by many (or all) genes in cells, cell types, tissues, or organs is called the RNA *expression profile*. Expression levels are determined by measuring the concentration of RNA molecules prepared from a biological sample by one of two means: (1) *microarray hybridization*, in which RNA concentration is inferred from the intensity of hybridization of the prepared samples on a patch of DNA complementary to the corresponding encoded RNA, or (2) by "deep" sequencing, in which RNA concentration is inferred by the number of times a stretch of sequence corresponding to an RNA molecule is detected.

DNA, RNA, and protein molecules can be acted upon by enzymes that add or remove chemical groups and therefore change their chemical properties. This process is called *enzymatic modification* of the molecule in question. Examples of enzymatic modification include methylation or demethylation of DNA or RNA bases, as well as phosphorylation, ubiquitylation, and glycosylation of amino acids that comprise proteins.

Box 1.3 Hormones, steroids, and pheromones.

Hormones are chemical signals synthesized by one cell type in an organism that modulate the physiology of another cell type in the same organism. Hormones can be peptides or proteins, in which case the receptor molecules for the hormone are present on the surface of the responding cell. Hormones can also be small molecules, in which case the receptor is intracellular. One example of a small molecule hormone is a steroid, which is a small molecule derived from cholesterol. Steroid molecules from one cell usually enter the target cells nucleus directly, and can modulate mRNA expression levels of the target cell through interaction with their cognate steroid receptor proteins present in the target cell nucleus. Some steroids are also structural and functional components of the cell membrane.

Pheromones are hormones that are synthesized by one organism but modulate the physiology of another organism.

Box 1.4) to the genome in the budding yeast. The specific pheromone we are interested in, an *alpha factor* pheromone, is a small protein that is received by the cell, and the cell responds to the pheromone by changing the expression levels of many genes. The pheromone does not actually enter the cell to cause its effects on the expression profile of the genome. Rather, it binds to a protein receptor molecule on the cell's surface, like a key to a lock. This physical "docking" of the pheromone to the receptor triggers an enzymatic activity of the receptor molecule, which extends into the interior of the cell. The activated receptor molecule then chemically modifies another protein, which in turn activates another protein and so on, until finally a protein that is a regulator of gene transcription is activated. This latter protein, upon activation, is able to enter the nucleus (which, when inactive, it cannot do) and bind to certain DNA sequences to bring about altered transcription at those locations (Figure 1.2).

Now each of these successive enzymatic modifications of the respective enzymes in the pathway is an individual chemical reaction that can be studied in great structural and chemical detail, which is important for understanding and thereby controlling individual steps (Box 1.4 introduces the concept of a signaling pathway). Figure 1.2 also shows that there are other signals in addition to the pheromone, such as high osmotic pressure, which share some, but not all, of the same intermedi-

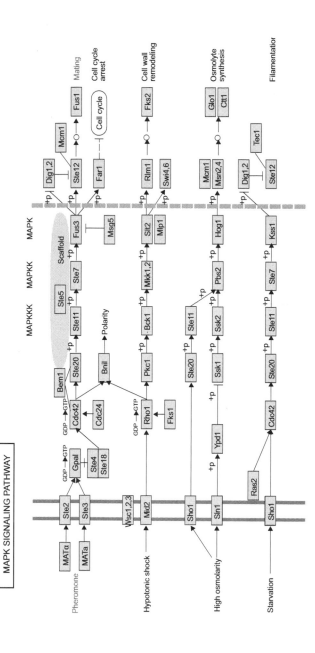

FIGURE 1.2: Signaling pathways in budding yeast. Rectangular boxes denote proteins, arrows denote positive regulation, lines with bars denote negative regulation. The solid gray vertical lines represent the cell membrane, while the dashed vertical line represents the nuclear membrane. Thus, the portion of the figure to the left of the cell membrane corresponds to the extracellular region, the portion between the cell and nuclear membranes is the cytoplasm, and the portion to the right of the nuclear membrane is the nucleus. Proteins embedded within the cell membrane are receptor proteins that bind and respond to external signals. These signals can either be chemical (e.g., pheromone protein MATa or MATα expressed during starvation) or physical (e.g., hypotonic shock; high osmolarity). $+p$ denotes a chemical reaction by which a phosphate group is added and $-p$ denotes one in which a phosphate group is removed. Reproduced with permission from Kyoto Encyclopedia of Genes and Genomes (KEGG; http://www.kegg.jp).

Box 1.4 Signal transduction or signaling.

The mechanism by which a chemical or physical signal is perceived, amplified, and carried to responder molecules within the cell such that a biological response is actuated is called a *signal transduction* or *signaling* mechanism. Signals may be physical in nature, such as light, osmotic pressure, shear force, and so on. Signals may also be chemical, such as hormones, pheromones, or neurotransmitters (chemicals that transmit information at nerve endings). Signals may be extracellular, such as hormones, osmotic pressure, or an antibody molecule or a virus that binds to the cell surface. They may be intracellular, such as an accidental DNA break in a chromosome within the nucleus, intracellular introduction of viral genetic material, or an abnormally folded protein within the cytoplasm. Signals are first detected by specialized receptor molecules. The specific binding or change in conformation of the receptor usually leads to a change in the enzymatic property of the receptor protein or one of the proteins that tightly binds to the receptor protein. This change triggers modulated expression of the product of the reaction catalyzed by the enzyme. Usually the receptor changes a property of one enzyme, which changes that of another enzyme, and so on, until the small effect of the original signal-receptor interaction event is amplified to a large change in the state (modification state) of some proteins or RNA or DNA in the cell. The pathway of successive enzymatic modification is called the signaling or signal transduction pathway. Because many of these pathway components interact, we increasingly speak of the networks formed by pathways and their interactions as *signaling networks*.

ate protein components and signal transduction pathways, and similar reaction mechanisms, to alter the level of gene expression of a different subset of genes in the genome. Competition among the two signaling agents, the alpha factor pheromone and high osmotic pressure, would therefore be expected to produce complex effects on the expression profile of the genome. Thus it is important to model these pathways as interacting networks, not just as independent reaction steps.

The same basic mechanism of signal transduction from the membrane-bound receptor protein exposed to the outside of the cell to the genome is also operative in human cells when they receive certain hormone signals, such as a hormone that signals to a cancer cell and the latter responds by increased proliferation. It is important to understand in great molecular detail each individual reaction for the purpose of developing chemical inhibitors of one or more of these reactions as an anti-cancer drug. However, the initial identification of these pathways, and later understanding of how these individual pathways interact with additional such pathways of signal transduction, are important on the one hand for identifying drug targets and on the other for predicting possible non-target effects (i.e., side effects) of a drug molecule.

We do not, of course, suggest that physico-chemical details of interactions are unimportant. In fact, such details are essential for testing hypotheses or to discover new molecular mechanisms. In that sense, they lead to fundamental understanding of biology as well as to successful applications of biological knowledge to practical ends, such as the discovery of new drugs or understanding disease mechanisms. Nonetheless, the higher-level abstraction of an interaction in the context of myriad others (at a low level of physico-chemical resolution) is essential in order to understand and appreciate the overall phenotypic consequences of the interaction in question. One of the themes of this book is to highlight our current ways of thinking about interactions in such a contextual manner.

Placing interactions within a contextual background often also helps reveal biological function. For example, a protein known as Microphthalmia-associated Transcription Factor (MITF) is important for converting certain skin cell precursors to the melanin pigment producing cells (called *melanocytes*) of the skin. The MITF protein is a transcription factor for a number of genes. Following the tradition of well-known transcription factor studies, it may be argued that it is important to study in atomic detail the interaction of this transcription factor protein with its cognate DNA binding site at each of the genes it regulates, and to explore the dynamics of its interaction at each of these genes. Ultimately this kind of study is extremely important, but it is immediately more relevant to determine the list of genes that it regulates, the list of additional proteins that might interact with MITF (which at least in principle might regulate the activity of MITF itself), and thus to generate a network of interaction. If one knows the biochemical function of some of its interaction partners, the relevance of the MITF protein becomes clearer in the context of additional functions. We thus understand that because MITF activates transcription of a gene that encodes a membrane channel protein called TRPM1, which is capable of sensing certain chemicals in the cell's exterior, melanocytes could have the ability to sense these chemicals. Furthermore, MITF also regulates genes that control cell division which are abnormally expressed in certain skin cancer cells. However, MITF is not produced at a high level in these cancer cells, suggesting that high levels of MITF control the rate of cell division. In sum, we infer from the interaction network around MITF that MITF regulates properties of normal melanocytes, including their ability to perform chemotaxis (movement in response to a chemical signal) as well as their propensity to participate in controlled cell division. This type of qualitative understanding is as important as understanding the atomic details of how MITF functions as a transcription factor. In the next section, we elaborate on specific examples of how the biological context influences our interpretation of the functions of genes and proteins.

1.1 Biological Function Depends on Context

Progress in the sequencing of genomes of a large number of organisms has made it possible to list with reasonable confidence nearly all protein-coding genes of these organisms. The next phase of progress in genomics depends on discovering the functions of these genes, and to develop models that should ultimately allow us to capture the phenotype from a description of the genotype.

In practice, the function of a gene is defined at several different levels. A geneticist describes the functional property of a gene in terms of the phenotypic variable that is lost or changed as a result of a mutation within that gene. For example, a rare mutant in fruit flies leads to the fly dropping down on its back and shaking its legs in an uncontrolled tremor when anesthetized. The corresponding gene is therefore termed the *shaker* gene, and its genetic property is defined as a gene that encodes a function required for normal maintenance of muscle tone and body posture. The *shaker* mutant flies are also defective in associative learning and memory, thus providing a mechanistic link between memory and behavior through the action of a single gene identified by a chance mutation. The *shaker* gene encodes a potassium channel protein, which suggests that muscle tone and associative learning are tantalizingly linked to the activity of this potassium channel protein that resides on the cell membrane.

It is often found that gene/protein properties are context dependent. In the previous example, the function of the *shaker* protein for maintaining muscle tone may be contextually irrelevant to its effect on associative memory in the brain. Let us consider a second example. A particular bacterial gene *recBCD*, when mutated, causes the bacterium to become hypersensitive to ultraviolet light and also unable to undergo sexual reproduction efficiently. These two properties of the *recBCD* gene's product are mechanistically related because the same biochemical activity of the gene product that allows repair of ultraviolet light-damaged DNA is utilized for incorporating DNA of the donor bacterium into the recipient bacterium following mating (Figure 1.3). The two properties are, however, revealed by totally different and apparently unrelated biological tests. The fact that the same gene mutation causes defects in the two biological activities provides a clue to the molecular mechanisms of both the repair of DNA damage and the ability of the organism to incorporate DNA in mating. A biochemist's description of a gene's function may include the chemical activity performed by the protein encoded by that gene (e.g., "potassium channel protein" for the *shaker* gene, or "DNA strand transfer" for the *recBCD* gene), but even that suffers from context dependence. For example, the function of the RecBCD protein can be described as the ability to unravel a DNA double helix into its two constituent single strands in one set of conditions as well as the ability to degrade the DNA into its constituent small molecular units under another set of conditions. These two different biochemical activities might superficially appear to be two separate chemical properties. Therefore, a complete understanding of the function of this gene must include not only the list of properties of the gene and its protein product from all relevant viewpoints, but also a list of the accompanying conditions that impose the context dependence.

An added layer of complexity in attributing function to a gene or its protein product is present because genes seldom carry out their role within a living cell as independent singletons. Rather, each gene or protein interacts in myriad ways with other genes or proteins, and each of these interactions is in principle required for the role that the gene has to play. Knowledge of the functional role of these other genes will therefore enrich our knowledge of the functional role of the gene in question. This inference is akin to inference of a person's role from the set of social or professional interactions that the person has with others. It is possible to infer personal attributes to a significant extent by knowing these interactions and, conversely, to infer putative interactions from personal attributes. In a similar vein, from a list of properties of genes

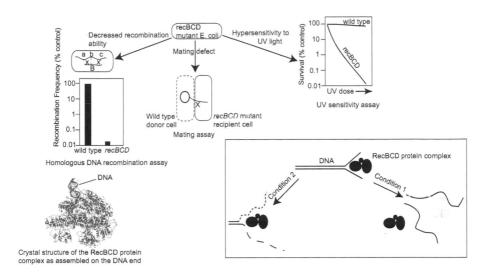

FIGURE 1.3: Context dependence of genes and phenotypes. *Escherichia coli* mutants in the *recBCD* genes exhibit defects in at least three different phenotypes: sensitivity to ultraviolet light (UV), the ability to transfer DNA by mating between a wild type donor strain and a mutant recipient strain, and in recombining two pieces of DNA sequences that are similar. These mutant phenotypes arise ultimately due to the loss of one or more biochemical functions of the 3-subunit RecBCD protein, which binds to both double- and single-stranded DNA, and its subsequent action on the DNA. Its action on the DNA is multifarious. In one specific physiological condition, RecBCD unwinds a double-stranded DNA into its constituent two single strands (this is a *DNA helicase* activity). In another condition, it digests double-stranded DNA on both strands processively, attacking it at one end and progressively digesting to the other end, with somewhat different specificities of chemical cleavage (shown here as small and large dashed strands). There are additional biochemical activities of this enzyme that are not shown here. However, many proteins with a diversity of biochemical functions are known, and it is quite likely that many more have not been discovered. This concept of context-dependent function is important to keep in mind when we consider the function of a gene or its encoded product. An interaction network containing the gene or protein in question could in principle provide a better perspective on function because it would reveal the interaction neighborhood: known functions of the network neighbors often provide clues to additional functional possibilities.

or their encoded proteins (which we will sometimes call *gene products*) and their *interaction* partners (collectively called the *interactome*), we can generate a richer view of functional description than possible by observing one gene at a time. This is a theme to which we will return at various points in this book.

1.2 The Nature of Interactions

At this point it is important to review what biologists mean by the word "interaction." Because all biological function may be ultimately traced to physical binding of molecules, biologists speak of physical interaction as the ability of one biologically relevant molecule to bind to another with reasonable thermodynamic stability. In practice, for such an interaction to reveal itself, it is also necessary that the interaction is strong enough to be detectable by at least one biological or chemical technique. Whether the interaction is biologically significant or not is irrelevant at this stage, because no physical technique has the power to answer the question of biological relevance. Ultimately, the biological relevance of a particular physical interaction must be demonstrated by genetic or cell biological means: the most common method is to abrogate the particular interaction in question and then to study the effect of this perturbation on cell physiology.

1.2.1 Protein–Protein Interactions

The existence of a protein-protein interaction implies that one protein is able to bind to another protein to form a protein complex that is stable enough so that the pair of interacting partners can be detected by some technique. There are several such techniques available for the detection of protein-protein interactions. These are briefly introduced here, and elaborated upon in Chapter 2. One common method is the *two-hybrid interaction*, in which the binding interaction between a pair of proteins can be detected inside the environment of a living cell (i.e., *in vivo*), albeit under somewhat artificial conditions (Figure 1.4a). A second method relies on the ability to pull down one protein by an agent (such as an antibody) that has high affinity for that specific protein from an aqueous solution, and then determining what other protein binds to the original protein and is therefore pulled down along with it. This is *co-immunoprecipitation* (co-IP) (Figure 1.4b). The two-hybrid

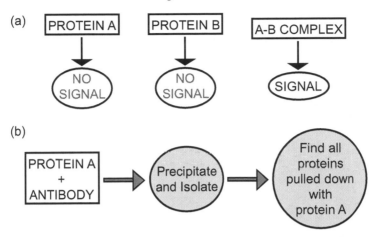

FIGURE 1.4: Two ways of discovering protein-protein interaction. (a) In the two-hybrid method, the coding sequence corresponding to protein A is fused to another protein that has the ability to bind a specific DNA sequence that is present adjacent to a signal gene (called a "reporter"), while the coding sequence corresponding to protein B is fused to yet another protein that has the ability to activate RNA synthesis. Neither of the two fused proteins alone is able to activate the synthesis of RNA corresponding to the reporter. However, if the two proteins are simultaneously synthesized and if the protein A physically binds to protein B, then the AB complex is able to bind the specific DNA sequence and activate synthesis of reporter RNA, which is then decoded into the reporter protein. The reporter protein is then detected by its catalytic activity. (b) In the co-IP method, a reagent that has highly selective affinity against protein A is added to a mixture of proteins containing protein A and many other proteins, some of which may or may not physically bind to protein A. The resulting precipitate is purified, and the mixture of proteins is analyzed to determine the identity of all proteins present in the mixture.

method only reports pair-wise interactions, although in reality some of the pair-wise interactions might be dependent on factors present within the cell that are invisible to the technique employed. By contrast, in co-IP more than two proteins are often simultaneously pulled down, which hides fine details of the direct interaction partners. For several organisms, lists of protein-protein interaction partners revealed by one or both of these methods are publicly available. The organism most comprehensively studied for these interactions is the baker's yeast *Saccharomyces cerevisiae*, in which protein-protein interaction maps covering nearly 6,000 encoded proteins are known.

Protein-protein interactions may signify a variety of biological functions. Two interacting proteins may do so merely to serve a tethering function. The proteins together may constitute a biochemically active molecule, such as an enzyme. Two proteins with distinct biochemical activities may be part of a multi-protein complex with a specific biochemical function. One protein might activate or inactivate the biochemical function of its partner by physical association.

1.2.2   Protein–DNA Interactions

A second kind of physical interaction is that between proteins and DNA. As the name implies, this physical interaction describes the binding of a protein to a specific DNA sequence motif. Examples include activators and repressors of transcription that bind specific DNA sequences tightly. As mentioned earlier, these proteins are generally known as transcription factors, and the specific DNA sequences to which they bind are termed *binding sites.*

A protein-DNA interaction is represented as a directional interaction from the protein molecule to the gene that is regulated by it. The direction of the interaction represents the direction of signaling by the protein to the expression state of the gene corresponding to the DNA in question. A protein-DNA interaction is identified by various types of experimental techniques. One of these depends on the ability of a protein to dock onto a DNA sequence and thus control a reporter gene activity that is measured. Reporter gene activity is thus triggered by the successful binding of the protein to the DNA sequence. Another technique isolates a complex of the test protein and the test DNA sequence, and identifies the complex (this method is called *Chromatin immunoprecipitation,* or ChIP). These methods are discussed in some detail in Chapter 2.

Protein-DNA interactions may also imply different kinds of biological function. For example, in some cases a protein may bind to a particular DNA sequence and thus activate or repress the synthesis of RNA from a nearby gene. In other cases a protein may bind to a specific DNA sequence and then cleave the DNA double helix.

1.2.3   Genetic Interactions

In addition to physical interactions, biologists admit another, more abstract kind of interaction—the so called *genetic interaction,* which may or may not represent a physical interaction between the partner genes or proteins.

One important kind of genetic interaction is the *synthetic lethal* in-

Box 1.5 Essentiality.

In genetics, *essential* genes are genes whose mutation causes complete loss of function of the gene product so that the organism cannot survive. Note, however, that some genes may be nonessential under laboratory conditions, but essential in the natural habitat, or vice versa. Therefore it is important to carefully consider the physical conditions under which a gene might be considered essential or nonessential. The definition of essentiality can become rather complex in multicellular organisms that have different cell types. For example, some genes are essential because no cell can survive when they are mutated. This is usually the case for genes that control core biological functions, such as DNA replication, RNA and protein synthesis, cell division, many aspects of energy utilization and metabolism, and transport of chemicals into or within the cells. However, there are other genes that can be essential only for a specialized cell type. Many such genes, if mutated, might support growth and development of the embryo only up to a degree, but later the embryos become defective and ultimately die. Such mutations are called *embryo lethal* mutations, and define a different type of genetic essentiality. Many inherited forms of human diseases are caused by mutations in genes whose normal functions are essential in only certain cell types during specific stages of development, often after birth.

teraction. Consider two mutations in two separate genes, such that neither mutation separately kills the cell, and yet the cell dies if both genes are simultaneously mutated. In this case the two genes are said to interact via a synthetic lethal interaction. Synthetic lethality of pairs of genes might represent different functional relationships in different situations. In one case, the two genes might redundantly control an essential biological process such that their simultaneous elimination leads to a loss of viability (see Box 1.5 for a discussion of essentiality). Alternatively, the loss of one gene might allow readjustment of the internal flow of biological information transfer within the cell such that a second "conduit" is now recruited to perform an essential process. Here the two gene functions are not redundant but are versatile. However, in the absence of both genes, the essential function is no longer salvageable and so the cell dies. Yet another example of genetic interaction is usually seen in the case where the mutation in the gene does not cause a complete loss of the encoded function but only partially destabilizes the affected gene product. Imagine that these two gene products are both needed for the same biological process. Here a single debilitating mutation might still allow just sufficient function for life, but two simultaneous mutations push the cell beyond recovery. Therefore, the interpretation of a particular synthetic lethal genetic interaction remains obscure in the absence of other information. Figure 1.5 shows various interpretations of synthetic lethal interactions that are possible given protein-protein interaction data. Note that synthetic lethal interactions, by definition, have no directionality. We revisit the experimental identification of genetic interactions, particularly synthetic lethal ones, in Chapter 2.

A second kind of genetic interaction is the *suppressor* interaction. An example of this type of interaction is the disruptive effect of a gene mutation that is compensated for by loss or gain of another gene's activity (i.e., modulation of the second gene's activity *suppresses* the disruption caused by the first gene). Consider two genes a and b whose encoded products, the proteins A and B, respectively, interact via protein-protein interaction (Figure 1.6). One interpretation of a suppressor interaction occurs when a mutation in gene a causes expression of the misfolded protein A^*, which has a defective interaction with B. Imagine that the defective interaction causes disruption in some biochemical process. Subsequently, a second mutation, this time in gene b, might cause misfolding of B to B^* in such a way as to make B^* regain the ability to interact with A^*, thus restoring biochemical activity. This type of genetic interaction is a direct proof of the existence of a physical interaction in a physiological context, as we discuss in the next chapter.

Another kind of suppressor interaction occurs when an essential biological process is eliminated by one mutation while its essentiality for life is removed by a second mutation (Figure 1.7). For example, one mutation may cause the absence of synthesis of one metabolite, but a suppressor mutation might cause the shunting of another metabolic intermediate to replace the missing substance by an alternate pathway. A third case of suppressor interaction may occur when one mutation causes the accumulation of a toxic substance and the second mutation removes it (or does not cause its accumulation) (Figure 1.8). Such suppressor interactions are also called *epistatic* interactions. Unlike synthetic lethal interactions, suppressor interactions are directional.

Clearly, genetic interactions are difficult to interpret in the absence of additional information on the nature of the genes and gene products involved in the process. It is also easily seen that the interpretation of

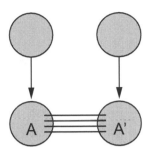

(a) Essential function

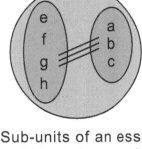

(b) Sub-units of an essential
multi-protein complex

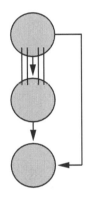

(c) Essential linear
pathway

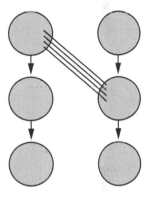

(d) Parallel pathways essential
for survival

FIGURE 1.5: Mechanisms of synthetic genetic interactions as proposed by Kaelin (2005). Each circle represents a set of mutually interacting proteins, that is, a multi-protein complex. Synthetic genetic interactions between corresponding genes are represented by solid undirected lines. Arrows represent signaling pathways (see Box 1.4). (a) Synthetic genetic interaction arising due to redundancy in an essential function; possibly the complexes A and A' have duplicate genes (paralogs). (b) Synthetic genetic interaction between two sub-units of an essential complex. (c) Synthetic genetic interactions between different components of a linear pathway—mutation in any one component causes partial loss of function, that is, reduced flow through the pathway. (d) Synthetic genetic interactions across two non-redundant but compensatory pathways. Adapted with permission from LeMeur and Gentleman (2008).

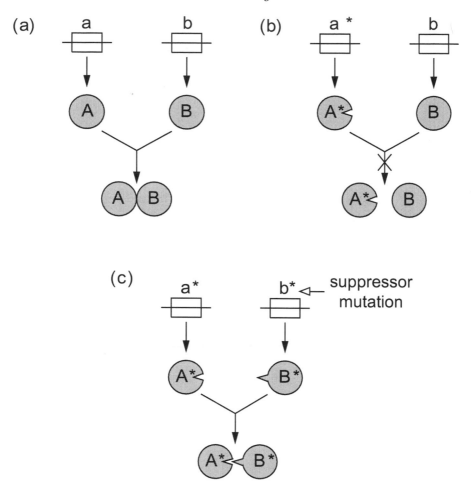

FIGURE 1.6: Mechanism of suppressor interaction by complementary mis-folding. Gene products A and B participate in protein-protein interaction. Mutation in gene a leads to gene a^*, which codes for misfolded protein A*. This effect can be suppressed by the mutated protein B*, which is misfolded in just the right way to restore the protein-protein interaction.

genetic interaction could in principle be aided by knowledge of physical interaction, and the existence of a physical interaction may attain functional relevance depending on whether or not a particular kind of genetic interaction also exists between the participants. These two types of interaction therefore yield complementary information.

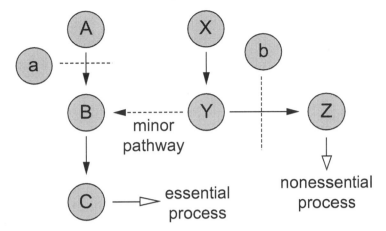

FIGURE 1.7: Mechanism of suppressor interaction by restoration of an essential process. Mutation in *a* causes disruption of production of B and therefore C. This effect can be suppressed by a mutation in *b* that prevents the production of Y to Z, leading to accumulation of increased levels of Y, which in turn leads to production of B by what is ordinarily a minor pathway. Increased levels of B restore production of C and therefore activate the essential process once again. Note that the process in which Z participates must be nonessential for this mechanism to work.

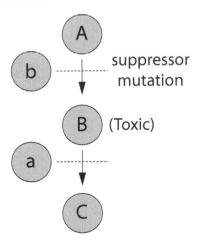

FIGURE 1.8: Mechanism of suppressor interaction by removal of toxic products. Mutation in *a* disrupts the pathway that converts B to C, thus leading to increased levels of B that are toxic to the cell. This effect can be suppressed by a mutation in *b* that disrupts the pathway for conversion of A to B.

1.3   Network as a Metaphor in Biology

A metaphor for understanding the functional relationships of genes and proteins in terms of their interactions, physical and genetic, is to compare individual proteins, genes, metabolites, and their functional associations with each other to the elementary building blocks and connections of an electrical circuit, where the fundamental units are electrical devices such as resistors, capacitors, inductors, and transistors.

The physics of the individual electrical devices is important in order to understand their characteristics and to engineer them. In a similar manner, detailed understanding of a single gene and its interactions with other biomolecules is important if we wish to re-engineer the regulation of the gene for possible therapeutic or bioengineering applications. The metaphor clearly illustrates, however, that in order to build and understand large computing architectures capable of specific computational tasks, it is relatively less important to study the physics of the individual electrical devices than it is to model the collective behavior of a system of many connected components.

Similarly, the objective of a new kind of biology that is becoming possible today, which is the focus of this book, is to understand the "functional architecture" of genes, proteins, and metabolites as they interact to produce a macroscopic phenotype. For this purpose, the detailed physics of the molecular architecture, the mechanisms by which the molecules interact and combine with one another, may be less important than their connectivity relations.

To achieve the goal of modeling and thus predicting the properties of a collection of genes and proteins, there are three separate overall needs, each achievable to varying degrees of approximation. The first is an exhaustive parts list, which is the list of genes, proteins, and their detailed biochemical, biophysical, and genetic properties. The second is an exhaustive list of interactions among these parts. This latter list may include purely qualitative information as well as quantitative information, such as binding constants, stoichiometry, concentrations, etc. These two kinds of information together allow one to draw a kind of wiring diagram for the cell. The third is a mathematical framework for deriving meaningful information out of the circuits. It is the wiring diagram itself that we refer to as the network, with the individual entities such as genes, proteins, or electrical devices playing the roles of *nodes* or *vertices*, and the connections between the entities referred to as *edges* or *links*. Thus, we speak of the protein-protein, the protein-DNA, or the synthetic lethal network, for instance. It is also possible to construct more detailed gene regulatory networks in which we can integrate the information obtained from individual interaction networks such that the integrated network represents more accurately the functional context of the network components. In the next chapter we discuss various methods, experimental and computational, for inferring networks from biological data.

1.4 Metabolic Networks

One of the most familiar networks in biology is the *metabolic network* or the network of metabolic pathways. Formally, this should be included in the category of physical interaction networks, but the metabolic network is significantly more complex than the protein-protein and protein-DNA interaction network so that it merits a separate discussion. Metabolic pathways denote the biochemical reaction routes by which chemical bonds are broken and/or synthesized for accomplishing three main kinds of biological function: acquiring and/or storing energy for accomplishing biological work, synthesizing or interconverting a variety of chemical compounds, and producing/transmitting biochemical signals.

In metabolic networks, the explicit nodes are chemical compounds, while the edges depict the direction of chemical transformation (Figure 1.9). In addition, there are implicit players in the metabolic network—the enzymes (mostly proteins) that bind to separate chemical entities (substrates) and transform them to other chemical entities (products). For (nearly) every chemical node, therefore, there is an enzyme that physically interacts with the substrate as well as the product. The chemical entities could be small, relatively simple molecules, or they could be complex molecules such as proteins or RNA. Table 1.1 lists some of the most common metabolites and the extent of their participation in metabolic reactions in the cell.

While formally every chemical reaction is reversible—this ensures the equilibrium balance between reactant and product concentrations—in practice most of the edges in biochemical pathways can be approximated as unidirectional directed edges. This is true because the series

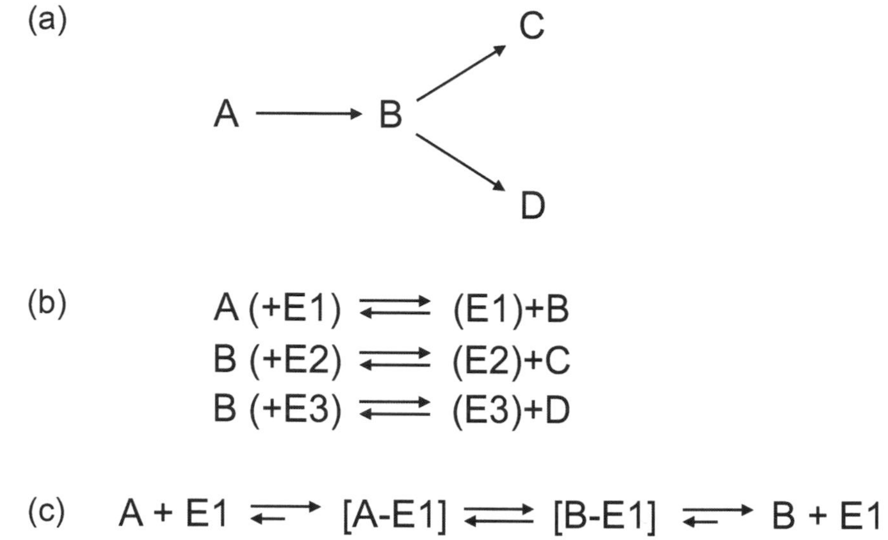

(a)

(b)

$$A\,(+E1) \rightleftharpoons (E1)+B$$
$$B\,(+E2) \rightleftharpoons (E2)+C$$
$$B\,(+E3) \rightleftharpoons (E3)+D$$

(c) $\quad A + E1 \rightleftharpoons [A\text{-}E1] \rightleftharpoons [B\text{-}E1] \rightleftharpoons B + E1$

FIGURE 1.9: (a) A simple metabolic network in which A, B, C, and D are four metabolite nodes, connected by directional edges, shows the pathways of conversion of A to C and D. (b) The series of enzymes (E1, E2, E3) drive individual reactions. (c) A simplified version of some of the typical physical interactions that occur at the first edge (A→B), in which the enzyme E1, having a high affinity for its substrate A, binds to A. This is an oversimplification, because in reality, A may have an unstable structural intermediate (called a transition-state intermediate, to which E1 might have a high affinity). A chemical transformation occurs in the complex A-E1 (or A's transition-state intermediate when complexed with E1) such that A is converted to B. E1 has low binding affinity for B, so the complex disassociates. Each binding reaction is governed by the equilibrium constant of that reaction.

of coupled enzymatic reactions in metabolic pathways, in which the product of one reaction is used as the substrate of the next reaction, with the final product being removed from the cell, usually maintains most of the reaction states far from equilibrium. Thus, one speaks of a *metabolic flux* that represents the flow of metabolites through the metabolic network. In Chapter 4 we discuss a particularly important analytical technique derived from the concept of metabolic flux.

Organisms differ somewhat with respect to the precise types of biochemical pathways that operate within their cells, although there are some core pathways conserved across organisms (Kanehisa et al., 2008;

TABLE 1.1: What Participates in Metabolism?

Metabolite	Degree	Metabolite	Degree
Proton	506	NAD^+	143
Water	390	Carbon dioxide	139
ATP	268	NADH	139
Diphosphate^{4-}	181	AMP	128
Phosphate^{3-}	178	Coenzyme A	119
ADP	173	Dioxygen	116
NADPH	166	Ammonium	92
$NADP^+$	166		

The most connected nodes and their numbers of connections (degree) in a yeast metabolic network in which nodes are metabolites and two metabolites are connected if they participate in a common reaction. *Source:* Herrgard et al. (2008).

Kuntzer et al., 2007). In Figure 1.10, the metabolic network that transforms carbohydrates, such as starch and sugars, into simpler chemical compounds, such as pyruvate, lactate, propionate, acetate, acetaldehyde, and ethanol, is summarized. The linear series of reactions starting from starch or sucrose at the top to pyruvate takes place in nearly all organisms. By contrast, the conversion of pyruvate to lactate, or to ethanol, may or may not occur in certain organisms, and when these reactions do occur in a particular organism, they represent specific evolutionary adaptations.

In general, biochemical pathways that accomplish synthetic tasks (making complex molecules from relatively simple ones) are termed *anabolic pathways*, while those that perform mostly degradative functions are called *catabolic pathways*, and the biochemical pathways that function mostly for generating stored forms of chemical free energy (energy that is available for accomplishing chemical, electrical, or mechanical work within the cell) are generally classified as *energy metabolism pathways*. Examples of catabolic pathways include the biochemical transformation routes for breaking down sugars and fat into carbon dioxide, water, and other small molecules such as acetate, lactate, or ethanol. Anabolic pathways include those that transform small molecules such as carbon dioxide, water, ammonium ion, and simple organic molecules (e.g., amino acids) into complex molecules such as proteins, polysaccha-

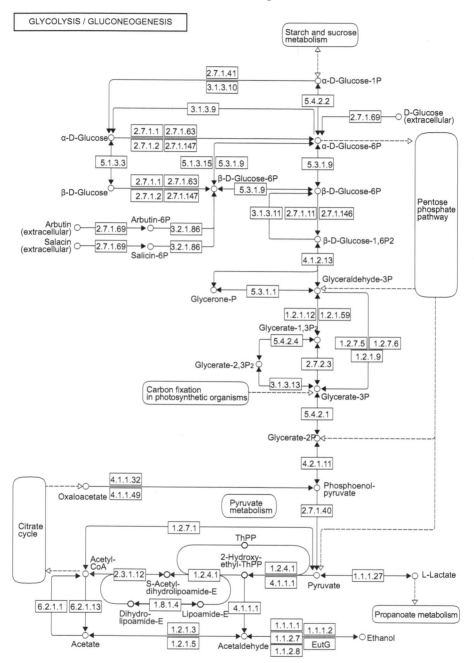

FIGURE 1.10: Metabolic pathways for breakdown of carbohydrates into simpler molecules. Each substrate or product is shown as an open circle with a chemical name identifier on its side. Each number within a box is an identifier [called the Enzyme Classification (EC) number] for the enzyme that catalyzes the interconversion of the substrate to the product. Boxes with rounded corners represent multiple reactions that are not shown in detail. Reproduced with permission from Kyoto Encyclopedia of Genes and Genomes (KEGG; http://www.kegg.jp).

rides (such as starch), and nucleic acids. Energy metabolism pathways include those of oxidative phosphorylation and electron transport chain in the mitochondria, oxidative pathways of fatty acid breakdown, and pathways that ferment sugar. It is important to understand that this classification is somewhat artificial because the three types of pathways are intricately intertwined. Thus, most if not all catabolic pathways are coupled to anabolic pathways, and both are coupled to energy metabolism.

A *signaling network* (see Box 1.4) is that subset of the metabolic and other physical interaction networks that has the special function of generating and transducing biological signals encoding certain informational states to which cells respond. One example is the pheromone signaling pathway alluded to earlier. In response to a communication between two cells (which might be mediated in a variety of ways, including the diffusive binding of an external protein molecule, the arrival of an electrical signal, direct chemical transformation on the cell surface, or entry into the cell of small molecules such as nitrous oxides and steroids), a series of chemical changes occurs within the cell. These changes may include the chemical breaking of a bond, causing the release of a small molecule from the cell membrane, a change in the three-dimensional structure or selective cleavage of an enzyme such that a new catalytic activity is induced, or the triggering of protein-protein or protein-nucleic acid binding in response to a small molecule. In this fashion, signaling pathways directly couple to metabolic, protein-protein, and protein-DNA networks [Lee et al. (2008b); Figure 1.11].

It is also important to appreciate that the metabolic and signaling networks are intrinsically part of the protein-protein and protein-DNA physical interaction pathways, because all enzymes are encoded by genes and most enzymes are proteins. Some (rare) enzymes are RNA molecules. Thus, genetic interaction networks are formally related to all physical networks. In the examples given in Figure 1.11, steroid biosynthesis in one cell type (specifically, in the adrenal gland) leads to transport of the steroid hormone through the blood into other cell types, such as the liver cells. In these cells, the steroid hormone enters the nucleus and binds to the cognate steroid receptor protein. The steroid-receptor complex then binds to certain specific DNA sequence elements (called *steroid response elements*) on the genome to alter mRNA synthesis activities of genes located next to the steroid response elements. Because there are multiple parallel pathways of sig-

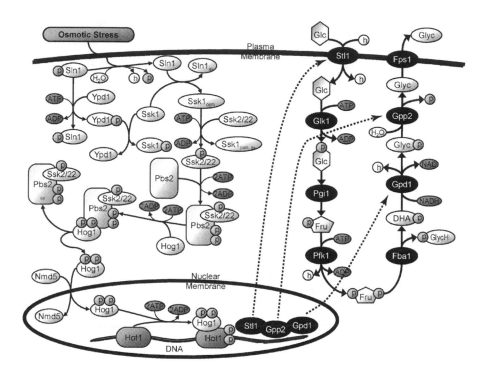

FIGURE 1.11: Coupling between two types of networks. The large connected network of proteins on the left is a signaling network that transduces a mechanical signal (osmotic stress due to high solute concentration inside the cell relative to the outside) from the cell membrane (the plasma membrane) to the genes located inside the nucleus. The mechanical signal (osmotic pressure) at the level of the membrane is amplified and transmitted through a cascade of coupled activating and inactivating system of enzymes into the nucleus through the activated Hog1 protein. The active Hog1 protein (shown with phosphate groups p) enters the nucleus, where it plays the role of a transcription factor: a protein that regulates transcription of its many target genes. Some of these target genes are Stl1, Gpp2, and Gpd1. The proteins encoded by these three genes catalyze the import of glucose (Glc) and the subsequent formation and efflux of glycerol (Glyc) (as shown in the unidirectional metabolic network on the right) from the cell cytoplasm into the extracellular space. This leads to higher glycerol (an osmotically active solute) concentration in the extracellular medium, causing an eventual reduction of osmotic stress by balancing the osmotic pressure across the plasma membrane. Dotted arrows signify the coupling between the signaling and the metabolic network. Reproduced with permission from Lee et al. (2008b).

naling and metabolism of steroids, mutations in the component genes that encode the metabolic enzymes and/or signaling pathways display the various genetic interaction types discussed above.

A remarkable finding to emerge after the whole genome sequences of a variety of organisms became available is the evolutionary asymmetry of gene family expansion by function. While the number of genes encoding the core biochemical and metabolic network components and those involved in essential functions, such as DNA, RNA or protein synthesis, and cell cycle control, have remained largely invariant across the evolutionary scale, genomes of different sizes display large differences in the number of genes involved in signal transduction and cell-cell communication (van Nimwegen, 2003). This observation suggests that an increase in the complexity of interconnections of interaction networks may provide a higher degree of physiological adjustability or robustness to the cell in response to a larger range of environmental fluctuations (van Nimwegen, 2003).

In other words, complex multicellular organisms appear to have evolved the capacity to sense and respond to a larger range of environmental signals by increasing the complexity of their networks and by acquiring the ability to respond to a wide variety of signals. The core network, undoubtedly, must have the capacity to readjust the flow of metabolites and energy in response to the larger range of signals. This might have been possible by fine-tuning the flow parameters as well as by generating novel connection topologies in the core network. However, no evidence in favor of either mechanism is yet available.

Alternatively, it has been suggested that symbiosis and/or mutualism among phylogenetically different organisms may have provided this additional capacity of metabolic core functions (Kitano and Oda, 2006). A loose association of populations of microbes, or *microbial flora*, for example, with multicellular organisms, may have given the host organisms a larger range of metabolic capacity without having to diversify their set of core genes. The important concept here is that an expansion of the host's gene family encoding signal processing components, and a correspondingly matched expansion of the symbiont's gene family producing signals, allow a broader degree of robustness. This theory makes sense of the observation that the bacterial microflora associated with an average human being consists of at least a thousand species, contributing approximately two million genes. The total amount of DNA contributed by non-eukaryotic symbiotic organisms is typically 90% of the total DNA recoverable from an entire human being (Kitano and Oda, 2006)!

Indeed, the intestinal microflora of the host are thought to contribute critical components of their mammalian host's immunity as well as physiology. These considerations bring to fore mechanisms of evolutionary adaptation of eukaryotic organisms by either adopting prokaryotic organisms as integral components of their genomes (as with the endo-symbiotic mitochondria and chloroplasts of Figure 1.1), but also by external symbiosis with microflora as a communal association with the host. Thus, networks of signaling between organisms connect the core networks within different organisms, which likely contribute to the physiological robustness of the entire living community to environmental fluctuations.

1.5   Robustness at the Network Level

A fundamental property of living systems is their ability to function robustly when faced with perturbations in their immediate environment. At the same time, however, living systems must also be flexible enough to adapt to different environmental conditions by changing their biological properties in a significant manner. These two seemingly contradictory features are, nevertheless, both essential for survival. We may therefore say that living organisms are "marginally robust," that is, they are only as robust as they need to be to survive in their immediate environment, but not so robust as to render them unfit to adapt to changing environments.

Marginal robustness seems apparent at the level of whole organisms, and has also been tested to some extent at the molecular level. For example, most proteins are only marginally thermodynamically stable. This ensures enough stability in order for them to adopt the structural conformations necessary for performing their functions, while at the same time provides enough flexibility so that a few mutations may allow a protein to adopt a new conformation corresponding to a new function.

The emergence of marginal stability over the course of evolution and, more generally, marginal robustness, can be understood as arising out of the alternating processes of mutation and selection: most mutations are known to be destabilizing at the protein level, leading to decreased organismal fitness. At the same time, natural selection ensures that the fittest individuals eventually dominate the population. These competing forces reach a delicate equilibrium at a marginally robust state.

While it is thus clear that robustness represents some fundamental property that is unaltered by perturbation and that, further, most biological systems should be marginally robust, we are still faced with a great deal of uncertainty as to what kind of perturbations the system must be robust to and what specific property is left unaltered. To an evolutionary biologist, this property is probably the organism's fitness as measured by its ability to reproduce. However, if we are to address robustness within the context of interaction networks, we face the question of how fitness can be measured as a property of the network itself. Because such questions do not, as yet, have a clear answer, different scientists use different overall properties as a proxy for fitness, and analyze the sensitivity of this overall property to different kinds of perturbations.

As one example, a classic perturbation experiment that can be carried out at the whole-genome level involves performing single-gene "knockouts" and consequently analyzing the organism's behavior.

A single-gene knockout is carried out by mutating or deleting part of a single gene in a manner so that it loses its function, either by a drastic decrease in its expression level, or by misfolding and subsequent rapid degradation of its protein product. The effect of this knockout perturbation can then be studied in terms of overall function. In the yeast *Saccharomyces cerevisiae* for example, approximately 20% of the genes are essential, meaning that knocking out any one of these genes results in cell death, while knocking out any one of the remaining 80% of genes does not appreciably alter cell growth. Does the fact that 80% of gene knockouts do not result in cell death imply that baker's yeast is a robust organism? The answer is not obvious. Many knockouts that do not kill the cell may alter its function in subtle ways. A large subset of these knockouts also slows down cell growth appreciably. At a higher level, however, the large proportion of genetic dispensability in yeast leads to the widespread notion that higher organisms are genetically robust, with the robustness being attributed to either (1) gene or molecular function duplication, (2) buffering or masking of a gene's function by an alternative biochemical pathway, or (3) involvement of certain genes in processes that are required only under extreme or untested environmental conditions. The first two notions of robustness are familiar to the reader from the earlier discussion of synthetic lethality, and indeed it may be argued that an organism with a dense synthetic lethal network is more robust than one that has a sparse synthetic lethal network.

One way to measure robustness at the network level is entirely computational. The method, however, relies on a somewhat artificial notion of robustness that does not have a direct experimental interpretation. A striking property of large biomolecular networks is that they possess a power-law degree distribution. As a case in point, the number of interactions of a protein in a protein-protein interaction network in baker's yeast is distributed as a power law, that is, the number of proteins that are connected to k other proteins is proportional to a power of k: $N(k) \propto k^{-\gamma}$ (Figure 1.12), where γ is called the *power-law exponent*. This implies that there are very few proteins that have a large number of interactions with other proteins (these are the so-called "hub" proteins), and that the majority of proteins have very few interactions with other proteins.

Imagine a computer experiment in which the hub proteins are systematically removed from the network one by one and the number of hub removals it takes for the network to fragment into disconnected sub-networks is measured. Such computer experiments have shown that protein-protein interaction networks are fragile (i.e., fragment easily) when a few hub proteins are systematically removed from them, but do not fragment easily when proteins are removed at random. This observation has led to the idea that protein-protein interaction networks, and more generally, power-law networks, are robust to random attack but very sensitive to directed attack on hubs. Note that the notion of "organismal fitness" here is formulated as something that depends on the structural integrity of the network: organisms with large connected networks are deemed more fit than those that have fragmented networks.

The computer experiment makes more biological sense once we also independently observe that hub proteins are highly enriched for essential proteins (Figure 1.12). Removing a hub protein is then akin to experimental knockout of an essential gene. Therefore, this computer analysis of robustness, carried out at the network level, would appear to be consistent with the results of the single-gene knockout experiment, at least insofar as the robustness idea is addressed.

Single-gene knockout analyses can be extended in obvious ways to

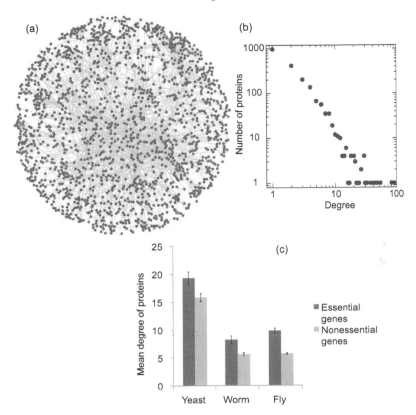

FIGURE 1.12: (a) A protein-protein interaction network in yeast, constructed using two-hybrid interaction data available as of November 2012. Each filled gray circle represents a protein, and edges represent protein-protein interactions as revealed by the two-hybrid method (see Chapter 2 for details of this method). Data was obtained from the BioGRID database [Stark et al. (2012); http://thebiogrid.org]. (b) Connectivity or degree distribution of interacting yeast proteins calculated from this data, giving the number of proteins (y-axis) that have a certain degree (x-axis). The connectivity distribution approximately obeys a power law with an exponent of about 2.25. (c) Proteins encoded by essential genes of several organisms are on average significantly more highly connected than proteins encoded by nonessential genes. Each filled bar represents the mean degree, while error bars represent standard errors. Data was obtained from Hahn and Kern (2005).

the analysis of effects of simultaneous knockout of multiple genes. Even finer analyses of robustness can be carried out if one has access to the detailed structure of biomolecular networks. For example, if the species concentrations of various proteins and metabolites as well as binding constants are known, it is possible to computationally simulate the effects of small changes in these concentrations and binding constants on the overall function of the network. It is not at all clear that results from such an analysis would provide the same view of robustness as knockout analysis would. Robustness at the network level may also be tested by studying the evolution of networks, that is, by comparing networks in related organisms and trying to ascertain to what extent the connectivities and network parameters are conserved. Again, it is not clear that robustness as revealed by conservation of these homologous networks would bear any resemblance to the other types of robustness discussed above.

Detailed accounts of experimental and computational methods to infer networks and test for their reliability are the subjects of Chapters 2 through 5. In Chapters 6 and 7, we discuss concrete models of small and large biological networks. In Chapter 8, we return to the theme of robustness, especially as it manifests itself in light of the modular structure of biological networks. Finally, in Chapter 9, we cover the role of networks, and more generally the "network paradigm," in illuminating disease mechanisms.

Chapter 2

Experimental Inference of Interactions

In this chapter we examine in some detail the experimental methods that are used to derive interaction networks. An interaction between two biological entities could be physical, for example an interaction between two proteins, between a protein and a length of DNA, or between a protein and a small molecule. An interaction could also be strictly informational, as in a genetic interaction. It is important to keep in mind that a physical interpretation of the latter is often difficult: a genetic interaction may or may not be due to a direct physical interaction, and often the basis of a genetic interaction remains abstract. Despite this difficulty, genetic interaction networks are considered vital to many important biological and medical goals. On the other hand, a physical interaction between two biological molecules can be simple to interpret but suffers from several uncertainties with respect to the relevance of the interaction for biological function. Thus, the two types of interactions—physical and genetic—often reveal complementary aspects of the biology of the organism.

2.1 Direct versus Indirect Inference of Interactions

Models of molecular or genetic interaction are based on two general types of experimental approaches: direct and indirect inference of interaction. Direct inference of interaction depends on an empirical demonstration of interaction between two entities without having to rely on an intervening model. All physical and genetic interaction data belong to this category. By contrast, indirect inference relies on a model or a hypothesis to infer interaction between a pair of biological entities.

Indirect inference may be of diverse types. In the case of interaction between pairs of proteins, for example, it may entail a computational search for the presence of protein sequence motifs among candi-

---

Box 2.1 Crystal structures of biopolymers.

Biological polymers, such as proteins and nucleic acids, can be crystallized under certain conditions, and the positions of the atoms that make up the molecule can be deduced from the diffraction pattern cast by the crystal when it is exposed to x-ray radiation. The conditions under which proteins crystallize can be rather artificial and drastic, often requiring highly concentrated solutions of protein in nonphysiological environments, such as abnormal salt concentrations and pH. Even so, it is remarkable that, by and large, crystal structures of biological macromolecules have been found to be consistent with the *in vivo* functional states, with few exceptions.

Just as a single macromolecule can be crystallized, it is possible to crystallize two interacting molecules, such as two proteins, or DNA bound to a protein, when the two molecules interact in a reasonably stable conformation. Structures found from a crystal composed of multiple interacting molecules are *co-crystal structures*.

---

date proteins. X-ray co-crystal structure determination of protein pairs (Box 2.1) have indicated that the interaction interfaces of most proteins have some conserved properties, such as largely hydrophobic amino acids (Box 2.2) buried into the core surrounded by partially accessible hydrophobic rings spanning an area of approximately 1,600 $Å^2$ per binding interface on average.

The interaction surfaces are usually composed of evolutionarily conserved protein domains (Box 2.3). Some of these domains, characterized by their amino acid sequences, have been shown by direct methods to physically interact when they occur in some specific proteins; yet, whether the presence of such domains would inevitably lead to protein-protein interaction is by no means certain, and a probability of interaction must be assigned in general. Theoretical and computational models that assign probabilities to whether a pair of proteins should interact can be constructed, and constitute examples of indirect inference. Such models are extensively discussed in Chapter 3.

A second common example of indirect inference is the prediction of DNA-protein interactions (also discussed in detail in Chapter 3) given detailed information on the tendency of the protein to bind to certain DNA sequences. One method is to examine a given DNA sequence for its similarity to experimentally observed DNA sequence(s) known to

Box 2.2 Classification of amino acids by their solubility in water.

The twenty naturally occurring amino acids that constitute proteins can be classified according to their chemical properties in various ways. A useful, although somewhat ambiguous, classification is based on hydrophobicity, or the extent to which an amino acid is soluble in water. Hydrophobicity is an important property because it plays a crucial role in shaping the three-dimensional structure of a protein: amino acids with hydrophobic chemical groups tend to be buried in the protein interior, where they are not exposed to the aqueous environment of the cell, while hydrophilic amino acids tend to be located in the solvent-exposed portion of a protein.

The extent of solubility of an amino acid in water depends on how strongly it interacts with water. Because water is polar (i.e., positive and negative charges in a water molecule are well separated), it tends to interact with other molecules that are charged or polar. On general grounds, therefore, one would expect amino acids that are charged or strongly polar to be hydrophilic. Thus, the negatively charged amino acids Aspartate (abbreviated as Asp or D) and Glutamate (Glu or E), as well as the positively charged amino acids Lysine (Lys or K) and Arginine (Arg or R), are hydrophilic. Serine (Ser or S), Threonine (Thr or T), Asparagine (Asn or N), and Glutamine (Gln or Q) are polar amino acids and therefore also hydrophilic. Nonpolar amino acids include Alanine (Ala or A), Valine (Val or V), Isoleucine (Ile or I), Leucine (Leu or L), Phenylalanine (Phe or F), Tryptophan (Trp or W), and Methionine (Met or M), which are hydrophobic. The polarity, and solubility in water, of the remaining five amino acids depend on their local environment within the protein. Histidine (His or H) and Cysteine (Cys or C) can be charged in many proteins in the cellular environment, and are hydrophilic in these situations, while when uncharged they are far less polar than many of the other polar amino acids. Tyrosine (Tyr or Y) has a hydroxyl group, suggesting it is water soluble, but that is offset by the presence of an aromatic ring which is typically nonreactive with water. Inside proteins, tyrosine can be either hydrophilic or hydrophobic. Glycine (Gly or G) has no side chain and usually behaves like a nonpolar, hydrophobic amino acid in proteins. Finally, Proline (Pro or P) usually behaves like a polar, hydrophilic amino acid in proteins.

bind to the protein in question. Here too, the inference of binding is indirect, and its accuracy depends on the validity of several simplifying assumptions.

Most methods of indirect inference are based on computational models. An example we discuss in Chapter 3 is that of gene regulatory net-

Box 2.3 Protein domains.

A protein domain is a contiguous part of the protein that may be an independently structurally stable unit, an evolutionarily conserved subsequence of the original protein sequence, or a single functional unit within a multifunctional protein. These three (rather loose) definitions have high overlap in practice. Some proteins are single domains in their entirety while many are multi-domain proteins. For example, all transcription factor proteins contain a *DNA-binding domain* that is responsible for attachment to specific DNA sequences; in addition, they may contain other domains that serve as binding sites for co-regulators and small molecules (ligands).

works deduced from successive expression of genes along a time series. Briefly, if a transcription factor gene TF is observed to be itself transcribed at time t_0 and a second gene A is observed to be transcribed at time t_1, where t_1 follows t_0, then TF is a candidate transcription factor for A (provided there is other evidence for TF regulating A), and we model the interaction pair as $TF \rightarrow A$. We see in Chapter 3 that indirect inference models can often be strengthened by additional information.

2.2   Physical versus Genetic Interactions

A physical interaction is equivalent to the "binding" of a biomolecule (protein or DNA) with another under approximate *in vivo* conditions. When present together in a cellular compartment at a certain temperature, A and B are said to physically bind when a complex AB exists in equilibrium with free forms A and B. The equilibrium concentration of the bound form relative to the free forms depends on the thermodynamic parameters of the binding reaction. Thus, all direct binding inference is ultimately subjective and depends on accepting a certain threshold concentration of the bound form AB as evidence of binding. Nearly all entities may bind to one another if the components are present in the same compartment in sufficient concentrations. The possible arbitrariness of this threshold concentration is the most important source of problems with physical interaction data: how strong must the binding be to be accepted as biologically relevant?

A "true" physical interaction implies the existence of at least some kind of genetic interaction, but the existence of a genetic interaction, depending on the type of genetic interaction under consideration, may or may not imply a physical interaction. We will see now that the existence of a particular type of genetic interaction can provide very strong support for the existence of a biologically relevant physical interaction in a manner that does not depend on a binding threshold.

Suppose that the equilibrium concentration of the bound form AB of two proteins is 1,000 times more than the product of the concentrations of the free proteins A and B, suggesting that indeed A physically interacts with B. Usually these parameters are measured *in vitro*, that is, under laboratory conditions rather than directly in the cellular environment. Therefore, the veracity of this interaction inside the cell under very different conditions is doubtful, because individual concentrations of A, B, and AB are uncertain in the cell and difficult to measure. Suppose, further, that there exists a mutation that happens to replace one or more amino acids of a protein A with some other amino acid(s) and that, as a result of this mutation, the mutated protein A* has a defective function. The defective function is usually expressed as a phenotype of the mutant gene *a*. Now imagine that there exists a second mutation in the gene encoding the protein B, also producing a different (unrelated) amino acid substitution. The mutated protein B* is itself defective, as shown by a mutant phenotype. Now, one of the following outcomes applies:

Compared to individuals with two normal (wild type) alleles (see Box 2.4 for a description of alleles) of the two genes (Table 2.1, case I), individuals carrying mutant alleles in either of the two genes are defective (Table 2.1, cases II and III). When individuals carry a mutant allele in each of the two genes, the most frequent outcome is that the individual will be defective (Table 2.1, case IV). However, in rare cases, such double mutant individuals will be normal rather than defective (Table 2.1, case V). When case V holds, we say that the mutational alleles in genes *a* and *b* suppress each other. However, it is possible that there are other alleles of the same genes *a* and/or *b* for which case IV, and not case V, applies. This type of genetic interaction, exhibited by case V, is known as *allele-specific suppression*—only specific combinations of mutational alleles in genes *a* and *b* show a genetic interaction (in this case, suppression of the defective phenotype). We have encountered this effect in Chapter 1 (Figure 1.6c). As discussed there, the only reasonable interpretation of allele-specific suppression is that proteins A and

Box 2.4 Alleles.

An allele is one mutant form of a gene. A gene can have many different alleles; for example, the DNA that encodes the gene might suffer mutations relative to the "normal" or "wild-type" allele of the gene. Alleles may correspond to varied phenotypes relative to the wild-type. They may be "silent" alleles, in which case there are no phenotypic consequences of the underlying mutation. Two different alleles may have identical or similar phenotypes. Alleles may represent a change in the amino acid sequence of the protein that the gene codes for—these are called *missense* mutant alleles. They may lead to a truncated protein because the underlying mutation leads to a stop codon in the middle of the original gene sequence. Alleles may correspond to complete deletion of the gene because of mutation in the transcription initiation region or the translation initiation region of the gene sequence. If the mutation occurs in the regulatory region, alleles may also entail production of more wild-type protein than normal.

TABLE 2.1: Allele Specific Suppression

Case	Gene A Allele	Gene B Allele	Phenotype
I	*Normal*	*Normal*	Normal
II	*Mutant-1*	*Normal*	Defective
III	*Normal*	*Mutant-2*	Defective
IV	*Mutant-1*	*Mutant-2*	Defective
V	*Mutant-1*	*Mutant-2*	Normal

B interact physically under normal (wild-type) conditions. The reasoning behind this interpretation is the following. Missense mutations in genes *a* and *b* cause misfolding of proteins A* and B*, respectively, by causing amino acid substitutions in the proteins. If proteins A and B must physically bind to each other for normal function, the misfolded protein A* may not bind properly to B, or misfolded B* may not bind to A, thus producing the defective complexes A*B or AB*, respectively. However, certain misfolded versions of A* and B* may have the property that the defects are exactly compensatory, and the two defective proteins are now able to bind to each other (A*B*), thus restoring biological function of the dimeric (or, more generally, multimeric) protein. The basic point is that it is very difficult to conceive of allele-specific suppression taking place without physical interaction between A and B.

The existence of allele-specific suppression is classically considered the most rigorous demonstration of the existence of protein-protein interaction. This may appear surprising because one might argue that there is no better demonstration of protein-protein interaction than the co-crystallization of two proteins revealed at atomic resolution. However, the issue is that biologists are mostly concerned not with whether two proteins can interact under *in vitro* conditions (such as the condition of co-crystallization), but whether they do so under normal biological conditions and whether their interaction has biological relevance.

As a case in point, despite the fact that two proteins will co-crystallize in a test tube, they might actually be confined to two different cellular compartments (e.g., one in the nucleus and another in the mitochondria) such that in a particular organism, the two proteins never have the opportunity to bind to each other. The fact that they can in fact bind might simply be an evolutionary relic: the binding might have had a functional significance in an ancestral organism but has none in the present organism. The existence of allele-specific suppression, on the other hand, guarantees the functional relevance of the particular interaction because biological function is expressed as a mutant phenotype or its suppression.

Genetic interaction can be of many different kinds, and each could have several different interpretations, some indicating direct physical interaction and others indicating the nonexistence of a physical interaction. We discuss some of these issues in a later section. We first examine a series of common methods for identifying direct physical interaction between pairs of proteins, which generates the major portion of existing data for interaction network analysis. While co-crystal structures of interacting proteins provide extremely reliable data on the possibility of two proteins interacting, such information is rare due to the difficulties in co-crystallization and structure determination. Nuclear magnetic resonance (NMR) structure determination also provides reliable data for the demonstration of protein-protein interaction, but like x-ray diffraction, this method cannot be rapidly scaled to produce large interaction datasets. Therefore, a number of fast methods for the determination of protein-protein interaction have been developed, and these techniques have revolutionized network-level studies in biology. The most common method for determining protein-protein interaction is the two-hybrid interaction conducted in a common experimental organism, baker's yeast.

2.3   The Two-Hybrid Interaction

The yeast two-hybrid (Y2H) method outputs a binary relationship between two proteins. One begins with the construction of two gene fusions. In one, the gene for one protein (X), also called a *bait*, is fused to the gene for a known DNA binding domain (DBD) that has the property to tightly bind a DNA sequence called the Upstream Activating Sequence (UAS) (Figure 2.1). This fused gene leads to the synthesis of a fused protein, denoted by DBD:X.

In the second gene fusion, the gene for a second protein (Y), also called a *prey*, is fused to the gene for another domain, called the activation domain (AD), which allows the synthesis of the second fusion protein, Y:AD.   The two genes are then introduced into the nucleus of yeast cells. DBD and AD fusions are so designed that the genes can be transcribed into mRNA at a very high level in the yeast nuclei, and the corresponding mRNAs are translated in the cytoplasm, following which the two fused proteins, DBD:X and Y:AD, are re-transported back into the nucleus.

A third gene, called a *reporter* gene, is also present in the nucleus. The protein product of the reporter gene is an enzyme whose presence can be determined by an appropriate chemical reaction or a phenotypic (e.g., growth) assay. The reporter gene has a few copies of UAS located close to the beginning of the gene. The AD present in the fused bait has the property to bind to an essential protein called a *mediator* whose function is to promote transcription of the reporter gene. However, the AD-mediator contact is quite unstable and, under experimental conditions, AD-mediator contact alone is insufficient to allow the mediator to promote transcription of the reporter gene. For transcription of the reporter gene to occur at a level that is sufficient to detect the enzymatic activity of its product, the AD-mediator complex must be stabilized in the vicinity of the reporter gene. The DBD:X fusion protein normally occupies UAS, but unless the bound protein interacts with Y:AD, the mediator is not activated. The reporter enzyme is synthesized and activated only when there is a sufficiently strong interaction between DBD:X and Y:AD (caused by the interaction between X and Y) such that the AD-mediator complex is stabilized. This is the signal accepted as indicative of a protein-protein interaction between X and Y.

The two gene fusions can be applied to many pairs of protein-coding

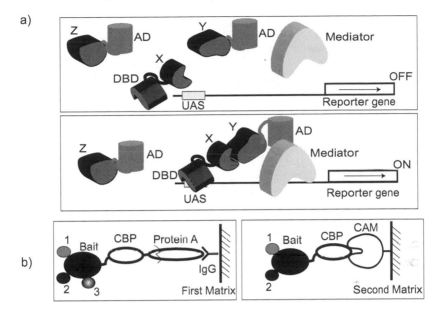

FIGURE 2.1: Two techniques for deciphering protein-protein interaction. (a) Two-hybrid interaction. The top box shows the relevant proteins not yet bound to the DNA. The arrow in the reporter gene corresponds to the direction of the reading frame. UAS is the upstream activator sequence for the reporter gene, AD is the activation domain, and DBD the DNA binding domain; proteins Z and Y are candidate interactors with X. Genes corresponding to the candidate interactors are constructed such that each gene is fused in frame to the gene encoding AD (examples are Y:AD and Z:AD). These fused genes are introduced into cells that also harbor a gene encoding a fusion between DBD and the bait protein X (DBD:X). For successful expression of the reporter gene to occur, the mediator protein must bind to the left of the reporter gene. In the top panel, the reporter gene is in the OFF state because the proteins are shown unbound to the DNA. In the lower panel, proteins X and Y are shown to be interacting, thus bringing AD in juxtaposition with the reporter gene and causing the mediator to bind upstream of the reporter gene. Reporter gene transcription can now occur, so that it is in the ON state. Thus, successful binding of X and Y is signaled by reporter gene expression. Once the cell that exhibits reporter gene expression has been detected, the identity of the gene that encodes Y is found by sequencing the DNA of the fused construct. (b) TAP tagging. CBP is a calmodulin binding protein and IgG is immunoglobulin G, an antibody; proteins 1, 2, and 3 are potential interactors with the bait protein. The left box shows the TAP tagged bait protein bound to the affinity matrix on which the IgG protein is immobilized. Protein A is subsequently released from the matrix by cleavage with an enzyme at a specific site indicated by a squiggly line. The mixture is then applied on a second matrix on which calmodulin (CAM) is immobilized (right panel). After washing the protein-matrix complex so as to remove proteins that bind to the bait protein with low affinity (in this case, protein 3), calcium ions are added to release the proteins from the second matrix. Subsequently, the proteins 1 and 2 that bind to the bait are identified by mass spectroscopy.

Box 2.5 Open reading frames.

The translation of a sequence of DNA or RNA nucleotides to a sequence of amino acids that make up a protein is carried out in the cell by translating each triplet of nucleotides, or codon, to the corresponding amino acid. A *reading frame* is a way of "reading" a DNA or RNA sequence in groups of three nucleotides at a time. There are thus three possible reading frames in an mRNA strand, each one starting at a different position in the sequence. An *open reading frame* (ORF) is a reading frame that begins with and contains the start codon ATG (in DNA) or AUG (in mRNA), and a subsequent stretch of sequence whose length is a multiple of three nucleotides but does not contain a stop codon (i.e., one of the codons TAA, TAG, and TGA in the DNA sequence, or UAA, UAG, and UGA in the mRNA sequence). An ORF therefore ends just before the first in-frame stop codon is encountered.

---

genes at a time, and pairs of interactions can be rapidly studied by robotically controlled introduction of the fused genes into yeast cells followed by automated detection of reporter gene activity. This method is highly suited to rapidly scan large genomes systematically. For example, open reading frames (ORFs, Box 2.5) corresponding to every gene in an organism can be synthesized and fused to *DBD* and *AD*, and all pairs scanned for reporter gene activity in yeast. In this fashion, nearly entire sets of the protein coding genes of yeast (*Saccharomyces cerevisiae*), fruit fly (*Drosophila melanogaster*), a nematode worm (*Caenorhabditis elegans*), and human have been scanned for the existence of protein-protein interaction (PPI). The first experimentally derived human PPI network was announced in 2005 by scanning for pair-wise interactions among 1,705 human proteins in yeast two-hybrid assays (Stelzl et al., 2005). Over 3,000 interaction pairs were identified in that work. Subsequent works using essentially similar techniques have, at the time of writing, amplified the human protein-protein interaction network to over 41,000 interactions involving over 10,000 proteins.

The normal human PPI network can be compared with the interaction network of proteins already linked to diseased states, such as cancers, as discussed further in Chapter 9. It was found that the connectivity properties of the PPI network of cancer-related proteins (corresponding to genes that are frequently mutated in cancer cells) is substantially different from the network of proteins hitherto not reported to be associated with cancers (Rambaldi et al., 2008). In particular, the cancer-related proteins have, on average, a large number of interactions as compared to other proteins in the human PPI (i.e., they have high degree in the human PPI). The differences are unlikely due to bias in the data (e.g., because cancer-related proteins are more exhaustively studied and therefore more of their interaction partners are known). This is because in comparing the cancer-related PPI network with PPI networks in several other organisms (yeast, fly or worm) where there is no cancer-related bias, the cancer-related proteins have on average significantly higher degree, more shortest paths connecting them, and higher betweenness centrality (Box 2.6) than non-cancer disease proteins, essential proteins, and proteins known to be important for controlling cellular events (Xia et al., 2011). The cancer PPI network is significantly more fragile than non-cancer or normal networks. Might these results suggest that cancer networks are more fragile because they have not been selected by evolution to robust properties?

Y2H data suffer from several sources of false positive and false negative errors. False positive errors may result from X and Y being produced in artificially high amounts, leading to high concentrations of their bound state, whereas *in vivo* they might not be present in high enough concentrations for the bound form to have significant concentration at equilibrium. Moreover, they are now forced to be present in the yeast nucleus whereas normally they may not both be expressed in the same intracellular compartment. Alternatively, the structure of at least one of the two fused proteins could be altered in the fused state in such a way that the interaction might be an artifact of the fusion itself.

False negative errors may arise from the possibility that the fused protein(s) cannot fold properly in yeast cells, cannot enter the nucleus, or are degraded rapidly in the nucleus.

Evidence indicates that most Y2H datasets have high false positive rates, while false negative errors are usually more frequent within certain classes of proteins such as membrane proteins. A manifestation of false positive errors in Y2H data is displayed by the only ~20% overlap between two independent genome-wide scans of the yeast genome, each of which discovered approximately similar numbers of interactions (692 and 841 interactions, respectively). Some techniques that rely on a very similar approach to Y2H, that is, by using protein fusions, but detect the physical interaction by directly monitoring close juxtaposition of the two fused proteins by optical techniques, such as fluorescence reso-

Box 2.6 Network centrality measures.

The positions and connectivities of nodes in a network can be mathematically described by various types of *centrality measures*. We have already encountered the simplest such measure, namely the *degree* of a node, which is simply the number of nodes that link directly to the node in question. Nodes may, however, occupy important, "central" positions in a network without having high degree. Imagine, for example, a node that is connected only to two other nodes (i.e., has degree 2), each having very high degree. In such a case, the "central" node is special not because it has high degree, but because all the paths that connect nodes lying at either side of the central node must pass through the central node. This example illustrates the importance of paths when trying to describe the centrality of a node. Of particular importance are *shortest paths* in a network: a shortest path between two nodes is a set of connected edges that links the two nodes in such a way as to traverse the smallest possible number of intermediate nodes. Many notions of node centrality rely on the concept of shortest paths. For example, consider a node v whose centrality we are interested in computing. Let $n_{st}(v)$ be the number of shortest paths that link distinct nodes s and t and pass through v. Let n_{st} be the total number of shortest paths that link nodes s and t. Then the *betweenness centrality* of v is defined in terms of the proportions of shortest paths between pairs of nodes that pass through v:

$$BC(v) = \sum_{s \neq t (\neq v)} \frac{n_{st}(v)}{n_{st}},$$

where the sum is over distinct nodes s and t, neither of which is identical to v. Other centrality measures are also based on shortest paths: the *closeness centrality* measure of a node is related to the arithmetic mean of the lengths of the shortest paths between the node in question and all other nodes, and the *information centrality* measure is related to the harmonic mean of the same lengths.

---

nance energy transfer (FRET), also suffer from nearly identical sources of false positive and false negative errors. This occurs because both of these techniques involve forced expression of fused proteins in the yeast nucleus. Despite these errors, Y2H interaction datasets form the most useful binary protein-protein interaction data available today. As we see in Chapters 3 and 5, by combining different sources of interaction data in meaningful ways, one can enhance the power of network analysis.

2.4 Affinity Co-purification

While the Y2H method provides information on binary interactions, affinity co-purification provides information on protein complexes. This is useful for the following reason: in an interaction matrix of three proteins A, B, and C, even if the Y2H method were to report all possible binary interactions, it is not certain that the ternary complex ABC exists. This could be due to the fact that one of the three proteins may not coexist in the same intracellular compartment simultaneously with the other two. Another possibility is that protein B interacts with A and C via the same surface patch and can therefore not interact with both simultaneously because of occlusion of A by C or vice versa. Conversely, in the absence of one of the three proteins, it is possible that any or all three binary interactions might be too unstable to produce a Y2H signal. In such a case, the Y2H data will have no information on the complex. To reveal information on higher-order interactions among proteins, several methods have been developed that can be grouped together as *affinity co-purification* methods.

The basic task in these methods is to purify a particular protein from cells and to detect the other proteins that are physically associated with it. This experiment is carried out under a set of stringent conditions such that most proteins that are unbound and weakly bound to the protein in question are eliminated from the mixture. There are two common ways to purify the proteins: co-immunoprecipitation (co-IP) and tandem affinity purification (TAP) by tagging.

In the co-IP technique, an antibody (Box 2.7) specific for one protein (the so-called "bait" or query protein) is used to affinity-purify it, and other proteins that are co-purified along with the bait are identified. Under these experimental conditions, it is customary to list the co-purified proteins as those that interact with the bait protein. This principle was discussed briefly in Chapter 1 (see also Figure 1.4a). The antibody may be specific to the whole protein, but in the more common scenario the protein itself is engineered to contain a domain that is recognized by an antibody (this process is called *epitope tagging*). The method is more reliable when the protein is engineered to contain two separate epitope tags and two successive immunoprecipitations are performed on the sample with the antibodies, thus reducing the number of accidental contaminants. This modified method is called TAP (or Tandem Affinity Purification) tagging. The method usually involves en-

Box 2.7 Antibodies.

When a foreign protein is injected into a mammal, the organism generally produces a type of protein that can bind with high specificity and selectivity to the injected foreign protein (called an *antigen*). This specific protein produced by the mammal as a response to the foreign protein is called an *antibody* protein. By virtue of their specificity and selectivity, antibodies can be useful "reagents" for the detection of the presence or absence of the antigen. Antibody molecules form the basis of the so-called *humoral immunity*—the defense mechanism of the host against invading organisms.

gineering the binding sites for two different proteins in tandem (protein A and CBP in Figure 2.1b). All proteins present in the final precipitate are then identified by mass spectrometry, which is a technique by which molecules are identified by their mass-to-charge ratios.

Protein A and calmodulin binding protein (CBP) shown in Figure 2.1b, whose binding sites are fused to the bait protein, have the property of tight (i.e., high-affinity) binding to their cognate binding sites. Protein A also binds tightly to a segment of the mammalian antibody protein IgG; this segment is called the "protein A binding site." The protein Calmodulin similarly binds with high affinity, in the absence of calcium ions, to CBP. In the presence of calcium ions, the affinity of Calmodulin to CBP is reduced. The fusion of the CBP and protein A binding sites to the bait protein is conveniently mediated by hooking up the DNA sequences encoding each of the two binding sites in tandem, in frame to the reading frame of the bait protein's DNA. Once the modified bait protein gene is constructed in this fashion, the fused bait protein can be produced in large quantities. Now this modified bait protein gene is introduced into a cell and the bait protein is expressed. Any protein (or any other molecule) that normally binds to the native bait protein is also expected to bind to the modified bait. It now remains to break open these cells and capture the bait protein on a matrix of beads or gel that has been pre-coated with IgG. IgG will bind to its cognate binding site on the modified bait, and the bait-interacting proteins will secondarily also stick to such beads. The beads are then gently washed to remove low-affinity binding proteins, and the bait is subsequently released from the beads by adding an enzyme that specifically cleaves the IgG protein off the beads. The effluents of the beads are collected and subsequently applied, in the absence of calcium ions, to another matrix that has been pre-coated with Calmodulin. The resulting complex mixture is again washed to remove low-affinity binding molecules. Subsequently, calcium ions are added to the mixture such that the bait protein, along with any high-affinity interacting protein that still remains attached to the bait, is released. These proteins are then identified by a variety of chemical techniques, such as mass spectroscopy.

Both types of affinity co-purification suffer from many sources of false positive and false negative errors. Accidental contaminants are the most common false positives, and accidental removal of genuine interactors lead to false negatives. Both methods rely on breaking open cells and thus mixing their internal content, which introduces another source of error due to disruption of the compartmentalization or localization of the constituent proteins. Additionally, the epitope or TAP tagged proteins are usually introduced into yeast or other cells by engineering their genes, and thus are usually expressed at artificially high levels, leading to the same sources of errors as in some Y2H experiments. Both Y2H and affinity co-purification suffer from the fact that post-translational modification in yeast is different from that in some other organisms: if the existence of interaction is dependent on a specific type of post-translational modification, then there will be errors in the detection of such interactions. Despite these limitations, affinity purification provides most of the protein complex data currently available for analysis. Combining Y2H and affinity co-purification data allows important insights into the network. For example, as we discuss in Chapter 5, intersections of multiple datasets can in principle provide high-confidence protein-protein interaction data, and help generate information about protein complexes.

We include in the class of affinity purification methods another type of analytical method, namely protein microarrays. Protein microarrays generate binary interaction data. In this method, proteins are separately printed in arrays on a coated surface such that the spatial location of each protein is well defined. Subsequently, a fluorescently tagged protein is applied over the entire microarray, and bound fluorescence at each location is monitored by a laser beam and recorded. While this method is extremely sensitive and very rapid, once all proteins are fluorescently tagged, it is dependent on the availability of pure proteins, which is an expensive and time-consuming step. The method is also sensitive to the physical conditions of binding, which is artificially produced on the arrays.

There is, however, an important ambiguity in the interpretation of affinity purification data: a protein that precipitates with the bait protein could do so either because it directly interacts with the bait protein or because it directly interacts with another protein that is precipitated with the bait protein. Thus we cannot unambiguously infer direct interactions between the bait protein and the individual proteins that are precipitated with it. There are, therefore, two extreme models for interpreting the results of affinity purification data: (1) the *spoke model*, which assumes that each protein that precipitates with the bait protein directly interacts with it but not with any other precipitated protein (leading to a "spoke-like" network structure); and (2) the *matrix model*, which assumes that *all* proteins that are precipitated (including the bait protein) directly interact with each other (leading to a dense, "matrix-like" network structure). The reality is usually somewhere between these two extreme assumptions but is impossible to tease apart using affinity purification data alone. Using yeast protein-protein interaction data culled from different types of experiments, Bader and Hogue (2002) found that the spoke model assumption is usually far more accurate than the matrix model assumption.

Because no single method for identifying physical interaction between proteins is reliable, it is imperative that network analysis is based on interaction data generated by at least two independent methods. A second consideration is that of the context of the reported interaction. If Y2H interactions are reported from small-scale experiments in which a single investigator applies a series of criteria to minimize experimental errors in a small dataset, then the resulting dataset is usually considered more reliable than Y2H data covering large numbers of protein pairs using "high-throughput" techniques (Box 2.8). Datasets present in a collective repository of protein interaction data (e.g., in BIOGRID; http://thebiogrid.org) are usually distinguished by the experimental context. Thus, a single report of Y2H interaction in a small-scale experiment is sometimes accepted as definitive proof for the existence or nonexistence of an interaction, whereas multiple reports from independent high-throughput experiments are necessary to reach the same level of confidence. Nonetheless, many of the same sources of false positive and negative errors specific to the techniques, not influenced by the experimental scale, may remain for small-scale experimental reports as well.

Box 2.8 High-throughput data in biology.

In biology, high-throughput data refer to data that are generated by a large number of experiments carried out simultaneously in an automated fashion. The experiments typically use robotic techniques and entail automated synthesis and analysis of the data using appropriate software. The classic staple of high-throughput biological experiments is a microtiter plate containing 96 "wells" arranged in a 8×12 rectangular grid. Each well is the site of an independent biological or chemical experiment. A typical high-throughput experiment involves many such plates and a robotic system to add and mix chemicals in each well, organize the plates, and read out the results. Modern microtiter plates contain larger multiples of 96 wells. At the time of writing, the "well" technology appears poised to be replaced by microfluidic technologies. For example, in a drop-based microfluidic technology, drops of fluid separated by oil replace the wells. Drop-based microfluidics offers the promise of faster screening using far less chemical reagent amounts. Because of their large scale and automated nature, high-throughput experiments are generally considered more error-prone than the same experiment carried out manually at a much smaller scale.

2.5 DNA-Protein Interactions

The second important kind of direct physical interaction is that between a protein and DNA. This information is essential for constructing models of gene regulatory networks. As mentioned in Chapter 1, a DNA-protein interaction network is a directed network in which an edge leads from a protein node to a gene node.

A number of small-scale experimental techniques can provide strong evidence for a protein binding to DNA. One such technique is an *in vivo* footprinting experiment coupled with a mutant study. In this method, cells are treated with a chemical that modifies DNA bases but is unable to modify a base if a protein physically occludes it (thus, the patch of unmodified DNA sequence constitutes a "footprint" of the protein on the DNA). The modified bases are identified following chemical treatment in normal cells and compared to those in mutant cells lacking the protein, thus leading to an identification of the footprint or binding site of the protein on the DNA.

One possible source of error here is that the protein may not itself bind the DNA but may indirectly control the binding of a second protein that actually occupies the DNA site. To eliminate this source of error, it is necessary to replace the amino acids of the candidate DNA binding protein and show that the footprints change in some regular manner.

Again, one of the best sources of evidence of DNA-protein interaction is a DNA-protein co-crystal structure, the determination of which is a time-consuming and difficult endeavor. As a result, high-throughput methods have been devised that are able to rapidly provide information on protein occupancy of the DNA, of which chromatin immunoprecipitation (ChIP) has been traditionally the most widely used technique.

In the ChIP method, one begins by making an antibody against the candidate protein or by tagging the protein with an epitope tag and expressing the protein in the nucleus of the organism in which the relevant interaction is interrogated. Cells in which the protein is expressed are treated with formaldehyde, a chemical that cross-links most proteins bound to the DNA to one another and also to the DNA bases. The cells are crushed, the protein-DNA complex (also called the *chromatin*) is purified away from the rest of the cellular material, and the chromatin is sheared into small fragments of approximately 200 to 500 nucleotides in size. These fragments are then applied to the antibody bound to a matrix, and the specific protein-DNA complex is separated from the unbound material. Subsequently, the chemical linkage between the protein and DNA is reversed, and the sequence identity of the DNA bound to the protein is revealed. This latter process is usually accomplished by labeling with a fluorescent molecule and hybridizing to known DNA sequences on a DNA microarray, or by direct sequencing. DNA microarray signal intensity, after being filtered for noise by averaging multiple measurements and, after subtraction of the background signal, serves as a good measure of the actual strength of the protein-DNA binding interaction.

Using the ChIP technique, all recognizable transcription factor binding sites over the entire genome have been determined for yeast. Genomic binding sites of a few chosen transcription factors have been determined for several other organisms, including fly, worm, and human.

Note that in the ChIP method, weak binding sites may remain undetected, and the resolution of the binding site (number of nucleotides bound) is determined by the average DNA fragment size at the immuno-affinity purification step. The latter factor is important for recognizing which gene corresponds to the DNA that is bound by the protein. In some cases, the distance between two genes might be relatively small compared with the resolution of the experiment in separating the boundaries between two adjacent protein-bound sites. In such cases, especially if the two genes are transcribed in opposite directions so that their respective regulatory regions both lie between the two genes, the exact identity of the gene that is actually controlled by the DNA binding protein may remain unresolved. The problem is particularly difficult in higher eukaryotes where a regulatory region controlling a particular gene can lie thousands or millions of base-pairs away from that gene.

A solution to the resolution problem is at least partially provided by the ChIP-SEQ method. As in the ChIP method, one also begins by immunoprecipitating the protein that is bound to the DNA (i.e., chemically cross-linked to the DNA at the binding sites). The immunoprecipitated or captured DNA samples are then sequenced. Obviously the sequencing must be carried out on the many thousands to millions of DNA molecules that are present in the sample. The sequencing technique provides single-molecule DNA sequences arising from clones of the sample DNA. The frequency with which a specific stretch of DNA nucleotides is found among the captured molecules relative to the frequency of that stretch in uncaptured DNA from the same sample (i.e., from un-cross-linked DNA) provides a measure of the extent to which the DNA target sequence is bound to the protein.

2.6   Classification and Interpretation of Genetic Interactions

Unlike physical interactions, genetic interactions are abstract, and each type of genetic interaction may be consistent with multiple alternative physical interactions. Given such lack of clarity, why do we study genetic interactions? The chief reason is that a genetic interaction, by the very fact that it is signaled by a phenotype, is of immediate biological relevance. Genetic interactions also enable us to organize biological information flow along defined, genetically controlled, pathways, to probe evolutionary processes, and to predict functions of unknown genes and proteins.

Studies of genetic interaction between two genes begin with mutated versions of each gene. Depending on its phenotype, a mutation could

TABLE 2.2:   Classification of Mutant Alleles of a Gene[1]

Allele	Deciding Configuration	Phenotype of Deciding Configuration	Classification
A	A/A	Normal (ϕ_A)	Wild-Type
a_1	a_1/A	Normal ($\phi_{a_1/A} = \phi_A$)	Recessive
a_2	a_2/A	Abnormal ($\phi_{a_2/A} \neq \phi_A$)	Dominant

be a reduction-of-function mutation, a complete-loss-of-function (null) mutation, or a gain-of-function mutation. Each such phenotypic class could be recessive or dominant relative to the normal (or wild-type) allele (Table 2.2).

We recognize two main classes of genetic interaction: *epistatic* and *synthetic*. Epistasis occurs when the double mutant *ab* in genes A and B have the same phenotype as that of one of the two single mutants (the logical or Boolean relationship between the phenotype of the double mutant and those of the single mutants is therefore $\phi_{ab} = \phi_a$ OR ϕ_b, where ϕ generically denotes the phenotype). Synthesis occurs when the phenotype of the double mutant is different from the phenotypes of either of the two single mutants. Synthetic lethality, a frequently used genetic interaction type, belongs to the second class. A finer classification of genetic interactions is given in Table 2.3 and illustrated in Figure 2.2. Note that some genetic interactions are undirected, whereas others are directed. On the other hand, in Figure 2.2 all interactions are directed (either positive or negative regulatory interactions). This is because while formally some interactions as deduced from genetic interaction data cannot be assigned a direction, in molecular terms genetic influence is always modeled as having some direction, such as the flow of a signal through a signaling network or the flow of metabolites through a metabolic network.

Let us consider the molecular interpretations of the genetic interactions listed in Table 2.3. It is most convenient for this discussion to examine interactions between genes whose phenotypes can be described quantitatively, such as the average height of individuals belonging to a genotypic class or the average activity of an enzyme for these individuals. For the "noninteractive" type of genetic interaction, the phenotype

---

[1]There exist other classes of gene mutations (null, semi-dominant, dominant negative, and haplo-insufficient) that require different gene configurations for testing. These are not relevant to the present discussion.

TABLE 2.3: Classification of Genetic Interactions[2]

Phenotypic Inequality	Classification	Interaction Representation
$\phi_{AB} = \phi_a < \phi_b = \phi_{ab}$	Noninteractive	A B
$\phi_{ab} < \phi_{AB} = \phi_a = \phi_b$	Synthetic	A – B
$\phi_{AB} < \phi_a = \phi_b = \phi_{ab}$	Asynthetic	A – B
$\phi_a = \phi_{ab} < \phi_{AB} > \phi_b$	Epistatic	A → B
$\phi_{AB} = \phi_a = \phi_{ab} > \phi_b$	Suppressive	A → B
$\phi_{ab} < \phi_a < \phi_b < \phi_{AB}$	Additive	A – B

See also Carter et al. (2009).

of Process II (Figure 2.2) is not affected by mutation in gene A, but is affected by mutation in gene b to the same extent regardless of whether gene A exists in wild-type form (as in Ab) or is mutated (as in ab). Although there could be a separate process (Process I in the figure) that is affected by mutation in gene A, the assignment of noninteractive is only based on the Process II phenotype.

In synthetic genetic interactions, the phenotype, which is jointly governed by the genes A and B, is not affected by either of the two single mutants (aB or Ab), but is affected by the double mutant (ab). One interpretation of the synthetic interaction is that the signaling process can travel through two separate pathways that ultimately converge to govern the process (see Figure 2.2). Thus, in the mutant aB, the signal finds the pathway through the node B; whereas in the mutant Ab, the signal is propagated through the node A. In the double mutant ab, the transmission of the signal is completely blocked, generating a synergistic phenotype.

In asynthetic interactions, the phenotype of the process governed by the genes A and B is affected to the same extent whether the mutant alleles are present singly or in combination. One interpretation of this phenomenon is that there is a third (unknown) gene (shown as an empty circle aligned with nodes A and B in the Asynthetic panel of Figure 2.2) that nonredundantly governs the same process.

In epistatic interactions, the phenotype of the double mutant ab is identical to that of one of the two single mutants (in Figure 2.2, the phenotype of ab is identical to that of aB). Note that an epistatic in-

---

[2]Additional classification based on phenotypic inequalities of the mutants and the wild-type is possible but will not be discussed here.

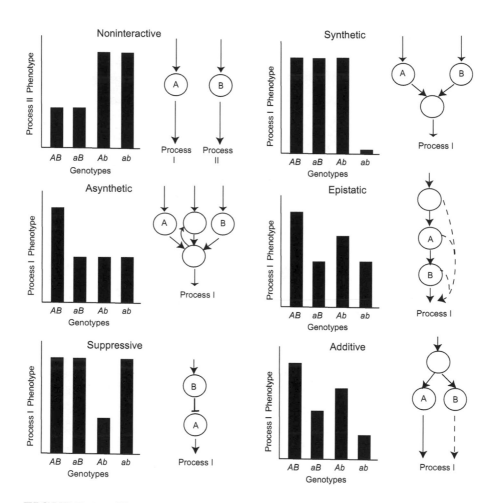

FIGURE 2.2: Illustration and interpretation of the genetic interaction types in Table 2.3. The magnitude of phenotypes corresponding to various genotypes are represented on an arbitrary scale. Genes or alleles labeled by italicized capital letters denote their wild-type versions (their protein products are represented by circles); those labeled by italicized lower-case letters denote mutants that do not express functional protein products. Thus, in a mutant *Ab*, the protein A is present but protein B is absent. Arrows represent the direction of signaling in a signaling pathway that is controlled by the genes.

teraction allows one to order the genes with respect to one another as well as provide a direction to the flow of signal in a signaling pathway. Such ordering by epistatic genetic interaction is also possible in a metabolic pathway (not shown here; see Figure 4.5) that, for example, is controlled by two enzymes A and B.

A suppressive genetic interaction occurs when the phenotype of one mutant allele (in Figure 2.2, that of *b*) is reversed (suppressed) by a second mutation (in this case, *a*). One interpretation of a suppressive interaction is that one of the two proteins that govern the signaling process (and thus control the phenotype) is an inhibitor of the other. In Figure 2.2, B inhibits A, such that under normal circumstances (as in the wild-type), the signal produces an absence of activity of protein A. In the mutant *aB*, the absence of the protein A is identical to the wild-type situation in the presence of the signal. In the mutant *Ab*, the signal is unable to activate B and thus activates A. As a result a mutant phenotype is exhibited. In the double mutant *ab*, however, the absence of the A function returns the system to a state that is effectively identical to that of *AB* or *aB*.

In the case of the additive genetic interaction, the effect of the signal is bifurcated into two separate processes, each contributing partially to the process. Here, mutations in genes *A* or *B* cause different degrees of phenotypic change; and in the double mutant, the effect of the simultaneous loss is additive.

An important point to note for all examples given in Figure 2.2 is that we assume that the signal is always present. More complex interpretations are needed when the phenotypes are measured under two different conditions of the signal state (ON or OFF). Moreover, there are several alternative interpretations and these also depend on whether the underlying process is a signaling or metabolic process [see the review by Avery and Wasserman (1992)].

It is interesting to note how genetic interactions can allow the construction of pathways of information flow through genes and their products. Take the case of two genes, *A* and *B*. The phenotypes of null mutations in either or both genes can help order the information flow through the two genes and determine network topology with respect to two other genes, *C* and *D*, that are already ordered (Figure 2.3). It is important to note here that such ordering of genes cannot be done unless the mutations in the genes are complete loss-of-function or null mutations.

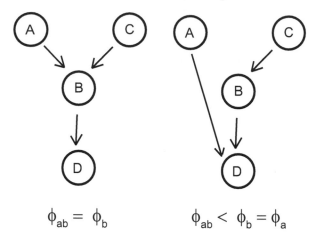

$$\phi_{ab} = \phi_b \qquad\qquad \phi_{ab} < \phi_b = \phi_a$$

FIGURE 2.3: Genetic interactions in signal transduction. Arrows denote the direction of signaling. In the left panel, mutation in gene B is epistatic to A because if there is a mutation in gene B the normal flow of information is blocked whether or not gene A is normal or mutated. In the right panel, the phenotype of the double mutant is more severe than either single mutant alone. In the context of signal transduction, proteins A and C may be two different receptor molecules that respond to the same type of signal, while B and D may be downstream proteins that amplify and transmit receptor activation further down the pathway. Similar arguments can be constructed for the genetic control of more complex biological processes, such as metabolic pathways. If we did not have knowledge of the true signal transduction pathway in the example above except for the fact that C is upstream of D, the pathway could be reconstructed using knowledge of ϕ_a, ϕ_b, and ϕ_{ab}.

2.7   Inference of Genetic Interactions

The essentiality of gene function for an organism's life under a particular environmental condition is indicative of the importance of the gene. Out of approximately 6,000 genes in baker's yeast (*Saccharomyces cerevisiae*), approximately 20% are essential, suggesting that in yeast each of the remaining approximately 80% genes is dispensable at least under laboratory conditions. Note that synthetic lethality is, in a sense, an extension of essentiality: it measures the dispensability of a pair of genes taken jointly, assuming that no single member of the pair is essential. It is therefore interesting to find the proportion of pairs of nonessential genes that is synthetically lethal. The density of the synthetic lethal genetic network, being related to the average dispensability of gene pairs, gives a finer picture of organismal robustness than that presented by essentiality data alone. To determine this network among nearly 12 million nonessential gene pairs in yeast, a technique has been developed in which arrays of single gene deletion mutant yeast cells are mated with all other single gene mutants and the resulting diploid cells are allowed to form haploid double mutants. The technique is termed the Synthetic Genetic Array (SGA) method (Tong et al., 2004).

The SGA technique is made possible by the biology of this organism. This particular species of yeast exists in three different cell types, called **a**, **alpha**, and **a/alpha**. The **a** and **alpha** cell types are haploid, which means that they have only one set of seventeen chromosomes (this configuration is denoted as $1n$). The **a** cells can mate only with **alpha** cells and vice versa. Upon mating, they produce diploid $(2n)$ **a/alpha** cells. The diploid **a/alpha** cells cannot normally mate with any other cell type, but can undergo a special type of cell division called *meiosis*, in which each diploid **a/alpha** cell can produce four haploid cells, two of which are of type **a** and two of type **alpha**. In this process, the two sets of seventeen chromosomes (which double in number because of one round of DNA replication) are equipartitioned into the four haploid cells, each of which gets one complement of the entire set of seventeen chromosomes.

If the **a** cell before mating had a mutation (denoted by a $-$ sign) in a nonessential gene 1, and the **alpha** cell had a mutation in the nonessential gene 2, then the **a/alpha** diploid cell produced by their mating would have one mutant and one wild-type copy of each of genes 1 and 2 (i.e., $1^-2^+/1^+2^-$). The four cells that will be produced upon meiosis would have a distribution of mutant and wild-type versions of the two genes (as expected from Mendel's law), and each cell would inherit exactly one copy of both genes 1 and 2. In other words, the four cells could be (1^-2^+), (1^-2^-), (1^+2^-), or (1^+2^+)[3]. If the two mutations were synthetically lethal, then the second genotype (1^-2^-) would be dead. If there is a way to select for the specific double mutant (e.g., by ensuring that each deletion mutation in the two query genes 1 and 2 is linked to a separate selectable marker gene, and by selecting for

---

[3]However, the relative frequencies of the four cell types need not be equal; the reader should convince herself why this is the case, and think about the conditions under which two of these cell types could be rather rare.

the presence of both marker genes in the cells after meiosis), then the double mutant would not be able to grow on the growth medium on which the double mutant should ordinarily grow. This failure of the double mutant combination to grow signals the presence of a synthetic lethal genetic interaction between genes 1 and 2. The ability to form a viable double mutant colony with growth rates that are similar to those of the parent single mutant strains shows the lack of synthetic genetic interaction. This technique is usually carried out in parallel on hundreds or thousands of deletion mutant pairs using robotic handling of cell cultures. In this manner, it is estimated that over 15% of the synthetic lethal interaction space has now been scanned, and the hope is that soon the entire list of synthetic lethal gene pairs will be determined for yeast.

The SGA technique outputs a binary value, corresponding to an unweighted synthetic lethal network, although the ability of the double mutant to grow poorly on a range of conditions is also captured, leading to an additional list of synthetic "sick" interactions (corresponding to doubly mutated cells that grow poorly rather than not at all) that is available from the BIOGRID database. Nonetheless, the output is still binary because no measure of the phenotypic intensity is directly provided. To obtain a measure of the phenotypic intensity (i.e., rate of cell growth) as a continuous value, a separate technique known as the Epistatic Miniarray Profiles (EMAP) technique is used. In EMAP, phenotypes (as measured, for example, by the rate of growth under a particular condition) of the single and double mutants are measured, recorded, and compared by image analysis (Schuldiner et al., 2005). In other words, the phenotypes are now quantified as a range of numbers, allowing full classification of the interactions as described in Table 2.3. This information produces a weighted network with directed edges. Networks derived from EMAP and similar efforts, where weighted genetic interactions can be produced, are likely to provide richer information than other currently available genetic interaction networks. In Chapter 4 (Section 4.3.1), we discuss a simple measure for interpreting the extent of genetic interaction quantitatively.

2.7.1   Finding Genetic Interactions with Inhibitory RNA

The elucidation of synthetic genetic interaction networks is a laborious task that has been possible on a large scale across the whole genome only in yeast, and recently in the bacterium *Escherichia coli*, due to

Box 2.9 Homozygous versus heterozygous organisms.

A diploid organism in which both copies of a gene have identical alleles is an organism that is said to be *homozygous* for the allele in question. If the two copies of the gene have different alleles, the organism is said to be *heterozygous* at the gene locus.

the existence of powerful genetic manipulation techniques for these organisms, as we have discussed. However, essentially the same principles can be applied to human cells for determining synthetic genetic interactions. The techniques, however, are somewhat modified in this case. In a recent study (Krastev et al., 2011), over 15,000 human genes were tested for synthetic genetic interaction with a gene called $TP53$. $TP53$ encodes an important transcription factor protein that is required for triggering programmed cell death when cellular DNA is damaged. The biological functions of $TP53$ (whose protein product is called p53) are accomplished in a number of different ways, chiefly through the modulation of the activities of several of its target genes, which encode proteins important for triggering cell death (see also Figure 5.1 and Section 5.2). The protein p53 is also important for activating the transcription of certain genes that are important for the repair of damaged DNA. In nearly 80% of all human cancers, the $TP53$ gene is mutated, suggesting that mutations in this gene predispose human cells for cancer. However, deletion of $TP53$ is nonlethal in mouse; furthermore, the homozygous deletion mutant cells in mice are not cancerous by themselves (see Box 2.9 for the distinction between organisms that are homozygous versus heterozygous for a certain allele). However, the homozygous $TP53$ deletion mutant mice are exceedingly *prone* to cancers. This suggests that mutations in other genes, which might accumulate in $TP53$ mutant cells, in conjunction with the loss of function mutation in $TP53$, might lead to cancer. Therefore it was thought important to query other genes for possible synthetic genetic interactions against a cell line derived from a homozygous $TP53$ mutant cell line.

A problem that arises in testing for synthetic genetic interactions in mammalian cells is that, unlike in yeast, mammalian cells cannot be coerced to mate or undergo meiosis in culture. As we discussed earlier, meiosis is necessary to generate the double mutant alleles in haploid cells, so that recessive alleles can exert their phenotypic effects. An indirect approach comes to the rescue. In the late 1990s, it was discovered that there exists a class of small RNA molecules, 22 to 24 nucleotides long, collectively called microRNA (miRNA), which have regulatory functions. These miRNA molecules are encoded by the genomes of nearly all eukaryotes (although not by *Saccharomyces cerevisiae*), including humans. The miRNA molecules do not encode proteins. Some miRNA molecules have sequences that are largely complementary to the protein-coding messenger RNA (mRNA) of other genes, especially to the sequences near the 3′ end region of the mRNA molecules. miRNA molecules can bind to their cognate target mRNA and thereby prevent the synthesis of proteins encoded by the target mRNA.

In certain organisms such as plants, miRNA molecules through complementary interaction can lead to the selective degradation of their target mRNAs, and also to selective blocking of further transcription of their corresponding genes. Whatever be the mechanism, the net effect of the miRNA molecule is the down-regulation of the expression of its target mRNA (or its corresponding gene). This effect of the miRNA molecule is often not very selective, because the sequence complementarity between the miRNA and the target mRNA molecules could be only approximate or because one miRNA might target multiple mRNAs (from separate genes) due to the fact that small stretches of complementary sequences could be present in multiple gene products. Nonetheless, miRNAs can be artificially synthesized and targeted against any gene (a technique that is termed *RNA inhibition* or *RNAi*), and can be used to shut off the expression of the targeted gene. The synthetic miRNA analogues are often termed *silencing RNA* or *siRNA*.

Returning to the *TP*53 story, synthetic genetic interactions with the *TP*53 gene were queried by challenging homozygous *TP*53 mutant human cells (derived from colorectal cancer cells) with artificial miRNA molecules raised against each of over 15,000 query genes. These homozygous TP53 mutant cells, each possessing one other down-regulated gene targeted by a miRNA molecule, were monitored in cell cultures through automated high-content time-lapse video microscopy to record cell phenotypes, such as division rate, growth rate, cell size, and cell shape. The phenotypes were compared with those of the parental cell lines (either the homozygous mutant cells without miRNA, or corresponding wild-type cells with the miRNA). This strategy allowed the identification of genes that need the function of wild-type *TP*53 for supporting normal growth as well as the identification of genes whose simultaneous loss with that of $TP53$ loss is important for normal growth. Out of over 15,000 genes so tested, 78 genes appeared to need the function of normal $TP53$, while the knockdown of the expression of only three genes in TP53 mutant cells caused drastic cell growth defects. Genes that need the function of wild-type $TP53$ gene appeared to be involved in the maturation and chemical modification of a number of different classes of RNA molecules, which appears to reveal an unanticipated role of p53. Synthetic genetic interaction approaches such as these are now increasingly engaged to unravel the molecular networks of interaction of genes important for human diseases (see Chapter 9).

2.7.2 Genetic Interactions in Organisms with Multiple Cell Types

So far we have only discussed genetic interaction networks through their combined phenotypic effects on single cells. Perhaps obviously, it is relatively straightforward to pose phenotypic questions on single cells—questions about cell division rates, cell size, or other physiological properties. However, understanding genetic interactions at the single-cell level does not necessarily advance our understanding of the impact of genetic interactions on phenotypes of complex organisms. The question of robustness of complex organisms having many different cell and tissue types and with complex developmental programs of cell specification is difficult if not impossible to frame in the context of genetic interactions. Yet these are important questions that must be addressed to fully understand the basis of biological evolution and function.

One of the earliest attempts to explore the importance of genetic interactions at the large scale (i.e., genome-wide systematic genetic interaction networks) for determining multicellular phenotypes used a nematode worm, *Caenorhabditis elegans*. This organism is tiny, with only 959 body cells and approximately 1,000 reproductive cells that undergo gametogenesis. It has a relatively streamlined genome with approximately 20,000 protein-coding genes and a short life cycle that renders the worm amenable to relatively rapid genetic manipulations.

Approximately 2,500 genes in *Caenorhabditis elegans* are essential for survival, and so far 554 genes are known to be essential for fertility. *C. elegans* is a transparent organism, and specific cell types in its gonad can be engineered to emit fluorescent light of different colors through genetic manipulation. These features make *C. elegans* a particularly

---

Box 2.10 The Pearson correlation coefficient.

Suppose we are given two lists of numbers, or two vectors, $\mathbf{v}$ and $\mathbf{w}$, with n entries in each list. The extent of correlation between $\mathbf{v}$ and $\mathbf{w}$ can be quantified using different types of correlation coefficients. A very commonly used coefficient is the *linear correlation coefficient*, also called the *Pearson correlation coefficient*. To compute this coefficient, one first computes the average or sample mean of the vectors $\mathbf{v}$ and $\mathbf{w}$. We denote these averages by $\bar{v}$ and $\bar{w}$, respectively. The formula for the Pearson coefficient is

$$r = \frac{\sum_{i=1}^{n} (v_i - \bar{v})(w_i - \bar{w})}{\sqrt{\sum_{i=1}^{n} (v_i - \bar{v})^2}\sqrt{\sum_{i=1}^{n} (w_i - \bar{w})^2}}.$$

where v_i and w_i denote individual entries in each vector. The Pearson correlation coefficient always takes a value between -1 and 1, with positive values signifying positive correlation and negative values signifying negative correlation or anti-correlation.

The Pearson correlation coefficient is called the linear correlation coefficient because of an interesting connection of this coefficient with linear regression analysis, which is the methodology for fitting data to a straight line. The square of the Pearson correlation coefficient is the proportion of variance in $\mathbf{v}$ that is explained by $\mathbf{w}$ (or equivalently, the proportion of variance in $\mathbf{w}$ that is explained by $\mathbf{v}$) as estimated by a straight-line fit of the graph of $\mathbf{v}$ versus $\mathbf{w}$.

It often occurs that two variables are found to be correlated because each variable is independently correlated with a third intermediate variable. In such cases, it is possible that the two variables in question would be uncorrelated if their correlation with the third variable could somehow be removed. The *residual* correlation between two variables when the controlling effect of a third variable is removed is called the *partial correlation coefficient*. Concretely, the partial correlation between X and Y after removal of the controlling effect of a third variable Z is given by the formula

$$r_{XY \cdot Z} = \frac{r_{XY} - r_{XZ}\, r_{YZ}}{\sqrt{1 - r_{XZ}^2}\sqrt{1 - r_{YZ}^2}},$$

where r_{XY}, r_{XZ}, and r_{YZ} are the Pearson correlation coefficients between X and Y, X and Z, and Y and Z, respectively.

special organism where the effects of gene pair knockouts on gonad structure and function can be relatively easily assessed.

Pairs of query genes from the set of 554 genes that are essential for fertility were systematically knocked down in a study by Green et al. (2011). The knockdown was effected by the addition of a pair of synthetic miRNAs, following which nearly ninety different cell biological parameters were measured by high-content, time-lapse video microscopy of the gonads. Such measurements conducted on the simultaneous knockdowns of nearly all gene pairs taken from 554 genes allowed the distinction of 102 separate phenotypic effect classes. Pairs of gene knockdowns that affected one or more of each of the 102 phenotypes were then determined. For each gene knockdown, therefore, one can mathematically construct a list with 102 entries, where each entry is a score that measures the influence of the knockdown on the phenotype. Similar lists can be constructed for each *pair* of knockdowns.

If the knockdown properties of a pair cannot be explained by the individual knockdown properties of the two genes separately, a genetic interaction between the two members of the pair is inferred, leading to the construction of a weighted gene interaction network. Furthermore, similarities among the phenotypic profiles of genes (as measured by, say, the Pearson correlation coefficient between the two knockdown lists, or vectors; see Box 2.10 for a definition of Pearson correlation) are indicative of functional similarity between the genes as evinced by the role they play in determining the structure and function of a complex organ in a multicellular organism. We anticipate that these types of approaches will increasingly provide a richer understanding of how gene networks determine complex phenotypes.

Chapter 3

Prediction of Physical Interactions

Theoretical and computational methodologies are indispensable approaches for the indirect inference of interactions, not just because they are more efficient and simpler to implement than experiments, but also because they often reveal the underlying principles of network connectivity and enable the prediction of new interactions in a systematic manner. Insights gained from successful computational models of networks can, in principle, be used to design new experiments that test these insights in a broad context.

It is very difficult to formulate a successful computational model for a biological network. In most cases, these models are based on assumptions that are too simple and therefore inaccurate. The relatively rare biological models that have few parameters and are based on general principles are called *unsupervised* models because their construction does not require detailed experimental data. Inaccuracies in these models are compensated by high-level insights into the processes being modeled. We will see examples of unsupervised models of large networks in Chapter 7. In many cases, however, it is important for models to be accurate in their details, and therefore to constantly confront them with results from experiments and tweak them to better capture these results. These biological models must have a more direct relationship with experiment than is prevalent in idealized models: they are constructed via a well-defined procedure that takes experimental data as input and outputs a set of model parameters. Such models are called *supervised* models because the numerical values of the parameters in them depend on the input data. The input data are called the *training set* for the supervised model. The training set changes as more data become available, leading to refinement in model parameters. The model parameters can in turn be used to predict new networks.

Supervised approaches that use properties of known network interactions to infer new network interactions fall under the category of *machine learning* approaches for network inference. While reasonably successful in terms of reproducing known interactions and predicting new ones, these models contain large numbers of parameters and are often too complex to grasp intuitively. The reason is a well-known theme in biology: simple models based on clear, general principles are often inaccurate because they do not capture the complexity of the details, while detailed models with large numbers of parameters fail to reveal general unifying principles. The challenge of finding simple yet accurate mathematical/computational models for large, complex biological systems remains largely insurmountable.

In this chapter, we begin our foray into computational methods for biological network inference. We start with a primer on entropy and information, because a basic understanding of these concepts is essential to understanding the computational inference of regulatory interactions. The primer is followed by a discussion of regulatory network inference, followed by a description of methodologies to infer protein-protein interactions.

3.1   Entropy and Information

Let us suppose that a gene X can be in one of several expression states, depending upon the external conditions. For the moment, we assume for simplicity that gene expression is Boolean; as discussed earlier, the genes in question can then be either ON or OFF, and we represent these two states by the letters O and F, respectively.

The concept of entropy (or, more correctly, Shannon entropy) aims to quantify the *uncertainty* in the expression state. Thus, for a gene that is always in state O or always in state F, the state is completely *certain*, and we would expect the entropy ascribed to the gene's expression level to be zero. On the other hand, if the state can be either O or F, we would expect the entropy to be non-zero because the state is uncertain. How can one quantify the extent of this uncertainty?

To be concrete, let us suppose that the gene in question is in state O with probability p and in state F with probability $1 - p$, and that we make a large number, say N, of independent observations of this state. We would then expect that, of these N observations, $m = Np$ find the gene in state O and $N - m = N(1 - p)$ find the gene in state F. The uncertainty in the expression state of the gene can then be quantified in terms of the number $\mathcal{N}(m, N)$ of different ways in which

m observations can be in state O, and $N - m$ observations can be in state F among a total of N observations [note that, if the gene were to always be in state O (or always in state F), the number of ways of writing a sequence of N Os (or N Fs) is just one]:

$$\mathcal{N}(m, N) = \frac{N!}{m!\,(N - m)!}. \tag{3.1}$$

The Shannon entropy of the expression level of the gene can be expressed as

$$H = \frac{\log \mathcal{N}(m, N)}{N} \tag{3.2}$$

when N is a large number of observations. The reason for the logarithm in the above definition is that it enables entropies of two independent expression states to be added to get the total entropy and also ensures that the entropy of a completely *certain* state is zero. The reason that the entropy is normalized by division by N, as will be clear in the simplification below, is to ensure that the resulting expression does not, in fact, depend on N. This is in line with the expectation that the uncertainty in expression level is an intrinsic property of the gene and should not depend on the number of observations of the state.

Let us now simplify Equation (3.2) to transform the entropy definition to a more familiar form. To do this, we make use of *Stirling's formula*, which states that $\log N! = N \log N - N$ for large N. Thus, the entropy formula can be simplified as follows:

$$H = \frac{1}{N} \left(\log N! - \log m! - \log(N - m)! \right),$$

$$= \frac{1}{N} \left(N \log N - N - m \log m + m \right.$$

$$\left. - (N - m) \log(N - m) + N - m \right). \tag{3.3}$$

Substituting $m = Np$ in the above leads to $\log m = \log N + \log p$ and $\log(N - m) = \log N + \log(1 - p)$, thus yielding the entropy function for a Boolean, or two-state, system:

$$H = -p \log p - (1 - p) \log(1 - p). \tag{3.4}$$

Note that the base of the logarithms in the above expression has not been specified. The choice of base is a matter of convenience; changing the base has the effect of multiplying the entropy by a constant factor.

When the logarithms are evaluated in base 2, we say we are measuring entropy in *bits*, or "binary digits."

Figure 3.1 gives a plot of the entropy of a two-state system, as measured in bits, as a function of p. As expected from a measure of uncertainty, the entropy is zero when $p = 0$ or $p = 1$ because the expression state is certain at these two extremes; furthermore, the entropy is maximal when the state is equally likely to be ON or OFF $(p = 1/2)$.

How do we define entropy for a system with more than two states? Let the number of expression states be s. Further suppose that X is in state i with probability p_i, where $i = 1, \ldots, s$. Thus, we can speak of the expression level probability distribution (or, simply, distribution) of a gene, which is the probability with which the gene is found in each expression state, across different time points in the cell cycle and across different experimental conditions. If the gene is always found in the same expression state (say, the high expression state) with high probability, then its expression level has low uncertainty; while if it is found with roughly equal probabilities in different expression states, then its expression level has high uncertainty.

The *Shannon entropy* of the expression level distribution of the gene is a simple generalization of the two-state entropy [for a complete exposition, see Shannon and Weaver (1963)] and is given by the formula

$$H(X) = -\sum_{i=1}^{s} p_i \log p_i. \tag{3.5}$$

As is evident from the formula above, if the gene is always found in a single expression state j with probability $p_j = 1$ (and all other probabilities zero), the Shannon entropy is zero, indicating that there is no uncertainty in the expression level of the gene. Furthermore, because the Shannon entropy formula depends only on probabilities of various occurrences, and not on the occurrences themselves, it is applicable not just to gene expression levels, but to any probability distribution.

Now let us suppose that we are interested in two genes, labeled X and Y. Because each gene is a 2-state system, the two-gene set is a $2 \times 2 = 4$-state system (in general, n genes would form a 2^n-state system), and the probabilities of occurrence for the four possible states are conveniently denoted as $p_{i,j}$, which represents the probability that gene X is in state i and gene Y is simultaneously in state j.

If the expression of genes X and Y were independent, then $p_{i,j} = p_i p_j$, but in general the two genes would not have independent expression

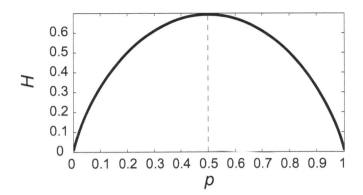

FIGURE 3.1: Shannon entropy of a 2-state gene expression model as a function of the probability p that the gene is OFF. The entropy is maximized at $p = 1/2$ and is zero when $p = 0$ or $p = 1$.

levels. We may then define the *joint entropy* of X and Y as

$$H(X,Y) = -\sum_{i,j} p_{i,j} \log p_{i,j}. \tag{3.6}$$

When X and Y are independent, the joint entropy can be expressed as the sum of the individual entropies: $H(X,Y) = H(X) + H(Y)$, because, in this case, $H(X,Y) = -\sum_{i,j} p_i\, p_j\, [\log p_i + \log p_j] = -\sum_i p_i \log p_i - \sum_j p_j \log p_j = H(X) + H(Y)$.[1] More generally, it is common to define the conditional entropies

$$H(X \mid Y) = H(X,Y) - H(Y),$$
$$H(Y \mid X) = H(X,Y) - H(X). \tag{3.7}$$

These conditional entropies represent the amount of uncertainty in one variable that is not shared with the other. For example, $H(X \mid Y)$ represents the uncertainty in X given knowledge of Y, or in other words, the residual uncertainty in X if Y were completely certain. When X and Y are independent, $H(X \mid Y) = H(X)$ and $H(Y \mid X) = H(Y)$, that is, the uncertainty in one variable does not change even if the other one is known with certainty—this is to be expected for independent variables. On the other hand, if X were completely determined by Y,

---

[1]In fact, it can be shown that $H(X,Y) = H(X) + H(Y)$ *only* when X and Y are independent.

the conditional entropy $H(X \mid Y)$ would be zero. As we shall see, these simple concepts play an important role when one tries to determine which genes regulate which other genes.

We now consider the shared or *mutual information* between X and Y, $M(X, Y)$, which is the amount of uncertainty in X if we remove the uncertainty associated with knowledge of Y (or equivalently, the amount of uncertainty in Y if we remove the uncertainty associated with knowledge of X):

$$M(X,Y) = H(X) - H(X \mid Y) = H(Y) - H(Y \mid X). \qquad (3.8)$$

Note that the mutual information is complementary to conditional entropy: it is zero if X and Y are independent, because, in this case, $H(X) = H(X \mid Y)$ and $H(Y) = H(Y \mid X)$. Alternatively, if Y completely determines X, the mutual information is the uncertainty in X. In general, the mutual information is always positive or zero and measures the extent of dependence between X and Y.

3.1.1   Interpretations of Mutual Information

The mutual information is a very important measure of the extent of dependence or correlation between two variables because of its statistical properties. Two of these properties that we now discuss—namely, the relationship between mutual information and the χ^2 probability distribution, and the meaning of mutual information as a likelihood function—are specially relevant to the inference of regulatory networks.

First, we note that the mutual information can be expressed as follows:

$$M(X,Y) = H(X) - H(X \mid Y) = H(X) + H(Y) - H(X,Y)$$
$$= \sum_{i,j} p_{i,j} \log \left[\frac{p_{i,j}}{p_i p_j} \right]. \qquad (3.9)$$

Returning to the gene expression example, the sum $\sum_{i,j}$ runs over the expression states of the two genes X and Y. Let us imagine that we make a large number N of simultaneous observations of the expression states of the two genes. In each observation, we record the expression state of each of the two genes. Let the number of times we find gene X in state i be N_i, the number of times we find gene Y in state j be N_j, and the number of times we find gene X in state i *and* gene Y in state j be N_{ij}. Clearly, $N = \sum_i N_i = \sum_j N_j = \sum_{i,j} N_{ij}$.

When calculating the mutual information, we estimate the probabilities in terms of the number of counts, as $p_{i,j} = N_{ij}/N$, $p_i = N_i/N$ and $p_j = N_j/N$. We now show that the corresponding estimate of mutual information can be expressed as a ratio of likelihoods. First, we define $q_{i,j}$ to be the true, unknown joint distribution of expression levels. Consider the following two hypotheses about $q_{i,j}$.

Hypothesis H_0: The expression levels of genes X and Y are independent; thus, $q_{i,j} = q_i q_j$.

Hypothesis H_1: The expression levels of genes X and Y are arbitrary.

In the language of statistics, the two hypotheses above are *nested hypotheses* because H_0 is a subset of H_1. The likelihood of the observed expression states under hypothesis H_0, which is the probability of the observed expression states calculated under the assumption that H_0 is true, is given by

$$L_0 = \prod_{i,j} [q_i q_j]^{N_{ij}} , \tag{3.10}$$

while the likelihood under hypothesis H_1 is

$$L_1 = \prod_{i,j} [q_{i,j}]^{N_{ij}} . \tag{3.11}$$

It can be shown that the above likelihoods reach their maximum value when q_i, q_j, and $q_{i,j}$ are equal to their values as estimated from the counts N_i, N_j, and N_{ij}, that is, when $q_i = N_i/N = p_i$, $q_j = N_j/N = p_j$, and $q_{i,j} = N_{ij}/N = p_{i,j}$. This yields

$$L_0^{\max} = \prod_{i,j} [p_i p_j]^{N_{ij}} , \qquad L_1^{\max} = \prod_{i,j} [p_{i,j}]^{N_{ij}} , \tag{3.12}$$

and therefore

$$\log \left[\frac{L_1^{\max}}{L_0^{\max}} \right] = \sum_{i,j} N_{i,j} \log \left[\frac{p_{i,j}}{p_i p_j} \right] = N M(X,Y). \tag{3.13}$$

This shows that the mutual information is proportional to the logarithm of a ratio of likelihoods of nested hypotheses: the higher the mutual information, the less does the data support the hypothesis of independence H_0 and the more does it support a general, correlated hypothesis H_1.

An important consequence of the meaning of mutual information in terms of a likelihood ratio occurs because of a standard theorem in statistics [see, for example, Miller (1955); Wilks (1962)], which states (under conditions that are satisfied for this example) that the probability distribution of minus 2 times the logarithm of a ratio of maximum likelihoods is the so-called χ^2 distribution when N is large (see Box 3.2). This is a particularly attractive mathematical property because it readily enables the computation of probabilities of occurrence of any range of values of the mutual information.

The two properties we have briefly outlined suggest why mutual information is a useful statistic for quantifying the extent of correlation between variables. We now turn to the question of inferring regulatory relationships among genes by using the mutual information between their expression levels.

3.2 Boolean Networks REVEAL Regulatory Interactions

In a regulatory interaction, one way that gene a can regulate the expression level of gene b is by coding for a transcription factor protein A that binds to the promoter region of gene b. The expression level of the transcription factor protein is in turn regulated by other transcription factor proteins. Such regulatory mechanisms are of fundamental importance because they dictate changes in gene expression level in the cell under different physiological conditions and across different tissues.

Microarray technologies enable the expression levels of a large number of genes to be simultaneously measured under different conditions. These conditions may also include different temporal stages in the life cycle of the cell, thus yielding the measurement of expression levels as a function of time. One of the foremost problems that could naturally be considered in the light of microarray experiments was that of inferring the gene regulatory network from time-course expression data.

A simple coarse-grained representation of expression level of a gene is to assume, as we did earlier, that the expression is *binary* or Boolean in nature, that is, a gene is either ON (has high expression) or OFF (has low expression). With this approximation, the expression states of an entire set of genes at a given time can be represented as a binary string of 0s and 1s, with genes that are OFF represented by 0 and genes that are ON represented by 1. By analyzing these binary strings at different time points and under different conditions, one can imagine trying to deduce what other genes completely "explain" the Boolean state of a given gene. For example, if a gene a is turned ON prior to another gene b being turned ON, gene a becomes a candidate explanatory gene for the expression state of gene b. As we shall see, things are naturally more complicated than that, but the relevant question here is: What Boolean expression states carry sufficient *information* to encapsulate the Boolean expression state of a given gene? And conversely, what Boolean expression states really do not add any relevant information to the problem? The answer hinges crucially on the concept of information, which we introduce below. If we can successfully identify the causative or explanatory genes for a given gene, these causative genes can be viewed as regulators of the gene in question, and so the Boolean network deduced from Boolean expression data may be interpreted as a regulatory network. Such Boolean regulatory networks are attractive from the point of view of biological interpretation, as they readily admit the representation of regulatory relationships as logical rules. A relatively simple example of these rules is provided by a small regulatory network that governs development of the mammalian cerebral cortex, which is a part of the brain that is highly developed in mammals and is required for information processing of higher cognitive function and volitional response. The regulatory network shown in Figure 3.2 involves five genes and is, to good approximation, a Boolean network with clear logical relationships between expression states.

We are now going to describe a relatively simple, information-based approach to the computational inference of Boolean regulatory networks from gene expression data that forms a basis for more complex techniques for reverse-engineering of the regulatory network. It was first implemented in the REVEAL (REVerse Engineering ALgorithm) algorithm proposed by Liang et al. (1998), building upon work by Fuhrman and Somogyi (1997) and Sawhill (1995).

3.2.1 The REVEAL Algorithm

The REVEAL method for identifying potential regulators of a gene G uses the concept of mutual information and proceeds iteratively. It first checks if any other *single* gene can completely explain the expression of G, that is, if another gene has the maximum possible mutual information with G. If no such gene can be found, it checks if there is a pair of genes such that the pair exhibits maximal mutual information with G, and so on. As this iteration proceeds, there is a combinato-

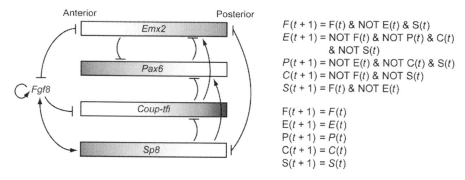

FIGURE 3.2: Five transcription factor-encoding genes implicated in development of the mammalian cerebral cortex. It is generally thought that the spatially restricted expression pattern of these transcription factors directs different patterns of target gene expression in the brain cells. As a result, these cells display heterogenous characteristics specified by their location in the brain. Expression of the *Fgf8* gene at the anterior pole of the cortex may be sufficient to drive the correct spatial patterning (shown in shades of gray) of expression of the transcription factors *Emx2, Pax6, Coup-tfi,* and *Sp8*. Arrows ($\rightarrow$) indicate activating interactions, while flat bars indicate repressive interactions. Boolean logic rules corresponding to the regulatory interactions are also shown, with F, E, P, C, S (in italics) representing the genes *Fgf8, Emx2, Pax6, Coup-tfi, Sp8*, respectively, and F,E,P,C,S (without italics) representing the respective proteins. For a gene to be ON at a certain time, its activators must be ON and repressors must be OFF at the previous time. Adapted with permission from Giacomantonio and Goodhill (2010).

rial explosion because of the need to exhaustively check all gene pairs, triplets, quadruplets, and so on. However, this combinatorial explosion is somewhat alleviated by the expectation that the expression levels of most genes will be "explained" by a small number of other genes. This is usually true for a large proportion of genes in an organism. Methods to alleviate the potential combinatorial explosion are discussed below, as are other generalizations of the REVEAL method. We first discuss some details of the basic method.

Consider the problem of finding potential regulators of G. We assume that we have access to expression data under different conditions so that there are multiple instances when G is OFF and ON. These multiple datasets are used to estimate the mutual information $M(G, X)$ between G and *every* other gene X. $M(G, X)$ is then compared to the entropy of the expression states of G, $H(G)$. If $M(G, X) = H(G)$, then X completely determines G, because the conditional entropy $H(G \mid X)$

is zero in this case, implying that the value of X has no variability once G is fixed. Thus the first iteration consists of finding the ratios $M(G, X)/H(G)$ and therefore identifying every gene G whose expression level is completely determined by some other gene X acting *singly*.

Now consider a gene G' whose regulator has not been found in the first step. That is, there is no gene acting singly that can explain the expression level distribution of G'. In the next step, the aim is to find potential regulators of genes like G' that have not been eliminated in the first step, that is, their expression levels cannot be determined by any other gene acting singly. We now consider all other genes acting *in pairs* and determine ratios of the form $M(G', [X, Y])/H(G')$, where $M(G', [X, Y]) = H(G') - H(G' \mid X, Y)$, and $H(G' \mid X, Y)$ is the conditional entropy of G' when X and Y are both completely certain. If the ratio is 1 for some pair X, Y, then this pair completely determines the expression level of G. In this manner, all genes that are explained by pairs of other genes are identified. The next step is to similarly identify genes whose expression is explicable by *triplets* of other genes, and so on. Note that in the transition from singlets to pairs to triplets and so on, the size of the search space is greatly increased; on the other hand, the number of genes whose regulators have not yet been identified decreases as the expression states of more and more genes are "explained." Also, because the mutual information to entropy ratios that need to be computed are all independent, the entire procedure is computationally parallelizable.

The REVEAL method thus infers, for every gene G, a set S of genes whose mutual information with G equals the uncertainty or entropy of G. These inferences can then be used to construct a directed Boolean network with edges directed from all members of S to G. This Boolean network may be interpreted as a regulatory network. Furthermore, the mechanism by which S regulates G is inferred by arranging the data in the form of a rule table that contains all possible states of S and the corresponding state of G that they explain (Figure 3.3). Liang et al. (1998) successfully tested the REVEAL algorithm on a set of 150 artificially constructed networks containing 50 genes each. In order to keep the problem computationally tractable, these networks were constructed so that each gene had at most three input edges (i.e., the expression level of each gene could be explained by the states of at most three other genes).

The REVEAL algorithm forms a core point of departure for various other related methods for inferring regulatory networks, some of which we describe below.

3.3   Variants of the REVEAL Method

The methodology described above for inferring Boolean networks can be trivially extended to multi-state situations, in which gene expression is taken to be in one of more than two states. This has the effect of making the problem more computationally expensive.

One limitation of the standard Boolean network methodology is its inability to infer self-feedback or auto-regulatory loops, which consist of genes that regulate their own expression. The reverse-engineering method always looks for other genes that explain the gene of interest, because within the framework of the method, the self-mutual information between a gene and itself is always maximal. If self-feedback were to be allowed within the methodology, REVEAL would always find that each gene is most capable of explaining itself, and therefore trivially (and incorrectly!) infer a regulatory network composed entirely of self-feedback loops. This property of REVEAL therefore appears to be a real problem unless one has access to *dynamic* gene expression data: expression levels of all genes in the set at different time points. Given such data, one can find the mutual information between a gene expression state at a particular time point and expression states of other genes at a *previous* time point. These other genes could include the gene of interest itself, and there is no reason why the mutual information between expression states of the same gene at different time points should be maximal (Figure 3.3). Thus, with dynamic expression data it is possible for the REVEAL method to infer self-feedback loops in regulatory systems in a nontrivial manner.

Another limitation of the standard REVEAL method is that it only infers deterministic regulatory mechanisms, because it looks for gene sets S such that the mutual information $M(G, S)/H(G) = 1$, or in other words, it looks for sets S that *completely* explain the variation in the expression of G. For example, the REVEAL method could infer that genes A and B in Table 3.1 regulate gene G via a mechanism based on the logical OR relationship. However, biological regulatory mechanisms are often subject to various sources of noise: intrinsic noise in the form of copy number fluctuations of regulatory protein molecules,

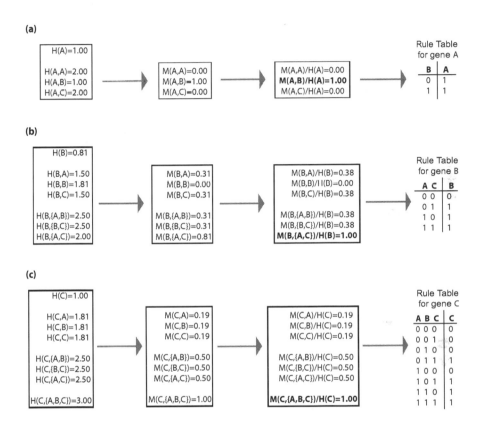

FIGURE 3.3: An example of the application of the REVEAL method (Liang et al., 1998). (a) A single gene B is sufficient to explain the expression level of gene A and is therefore predicted to be a regulator of gene A. (b) Genes A and C jointly regulate gene B. (c) All three genes—A, B, and C—are predicted to be regulators of C. In each calculation, mutual information is always computed between the expression levels of the gene under consideration and the expression level of every gene (including itself) *at the previous time point*. For all three cases, the expression data of the regulator gene(s) and the regulated gene are used to construct the rule tables.

TABLE 3.1: OR and NoisyOR Boolean Relationships

A	B	C = A OR B	C = A noisyOR B $\Pr(C = 0)$	$\Pr(C = 1)$
0	0	0	1.0	0.0
0	1	1	ϵ_1	$1 - \epsilon_1$
1	0	1	ϵ_2	$1 - \epsilon_2$
1	1	1	$\epsilon_1\epsilon_2$	$1 - \epsilon_1\epsilon_2$

extrinsic noise arising out of interactions with other molecules in the intracellular/nuclear environment, and finally, artificial noise arising out of modeling approximations, such as the 2-state approximation made in characterizing gene expression. It is therefore entirely plausible that regulators of a given gene do not deterministically explain the expression state of that gene but only do so in a *stochastic* or noisy manner. This type of relationship is exemplified by the noisyOR relationship in Table 3.1, where A and B together *nearly* determine the state of G via an OR relationship, but do not quite do so because of the presence of noise. Stochastic regulatory mechanisms are best described mathematically by *conditional probability distributions* that represent, for instance, the probability $P(G \mid A, B)$ that a gene G is in a certain expression state given the expression states of putative regulators A and B.

A fast variant of the REVEAL method, called ARACNE [Algorithm for the Reconstruction of Accurate Cellular Networks; Margolin et al. (2006)] is based on an extension of the concept of *relevance networks*. It involves computing mutual information values between all transcription factor (TF)-gene pairs, followed by a filtering step in which only pairs that have mutual information higher than some threshold value are retained. This threshold value is determined based on the probability distribution of mutual information scores of all TF-gene pairs pooled together. Following this, ARACNE implements a method to further filter out correlations that occur due to intermediate correlation with a third gene. This method uses the so-called Data Processing Inequality (DPI), which, in this context, states that if genes x and y interact only through a third gene z, then the pairwise mutual information satisfies the inequality $M(x, z) \leq \min [M(x, y), M(y, z)]$. Thus, interactions that pass through the first filter are collected into gene triplets. For each triplet, the edge with the smallest mutual information value is removed. The resultant network is then predicted to be the regulatory network. The method is fast because it only involves computation of pairwise mutual information, and there is no combinatorial explosion. It can therefore be readily applied at the genome scale. ARACNE has been successfully used to predict regulatory networks in human B cells (Basso et al., 2005).

Another method that is similar to ARACNE is the so-called Context Likelihood of Relatedness (CLR) method (Faith et al., 2007). The main difference between CLR and ARACNE is that CLR effectively uses different mutual information thresholds for each TF-gene pair to predict if interaction, and the thresholds are determined in a statistical manner. For each TF-gene pair, CLR pools together mutual information values of all pairs that contain either the TF or the target gene of interest. We call the probability distribution of these mutual information values the *background distribution*, which is used to assign an *empirical P-value* for interaction between the TF and the target gene of interest, as follows. Suppose that the mutual information of the TF-gene pair of interest is m^*. Using the background distribution, we can compute the probability that TF-gene pairs containing either the TF of interest or the target gene of interest will have mutual information exceeding m^*. This probability is the empirical P-value. TF-target gene pairs that have high P-values are considered similar to the background and are therefore not significant; low probability pairs have a significantly higher mutual information score than the background and are predicted to be regulatory interactions. This *context-based* approach has the advantage that it automatically filters out putative regulatory interactions by using data that is relevant to the TF and the target gene of interest. When applied to *Escherichia coli* genome-wide expression data, the method recovers known regulatory interactions at a better rate than other methods that are also fast (including ARACNE). The method also predicted a number of novel regulatory pathways that were subsequently verified experimentally.

Finally, the REVEAL methodology can be extended to optimize mathematical functions other than mutual information. In other words, a general mathematical function for the assignment of a specific set of regulators to a gene can be used in lieu of the mutual information. Such a function is called a *scoring function* because it assigns a "score" to each pair of genes; pairs that have a high score are predicted to interact, while pairs that have a low score are predicted to be noninteracting. The need for a scoring function other than mutual information is discussed in the next section. First, we point out that all of the variants and generalizations of the Boolean network methodology—multiple expression states, inference based on dynamic data, incorporation of noise, and general scoring schemes—can be addressed within the framework of *dynamic Bayesian networks*, which represent a powerful and flexible framework for the inference and representation of gene regulatory systems.

3.4   Dynamic Bayesian Networks

A Bayesian network is a directed network in which nodes represent arbitrary random variables (such as gene expression states) and edges represent conditional probability relationships between nodes and their parents.[2] In a Bayesian network, every node is independent of all other nodes given its parent nodes. Another way of thinking about this is that the probability distribution of the random variable representing a node in a Bayesian network depends only on values taken by the random variables representing the parent nodes. In the Bayesian network of Figure 3.4, Y_1 has the single parent X_1, and therefore the probability distribution of Y_1 is independent of the distributions of other random variables given the value of X_1. We thus have, for example, the following relationships: $P(Y_1, Z_1 \mid X_1) = P(Y_1 \mid X_1)P(Z_1 \mid X_1)$ and $P(Y_1 \mid X_1, X_2) = P(Y_1 \mid X_1)$.

A *dynamic Bayesian network* (DBN) is a special type of Bayesian network that represents the time evolution of a random variable or a set of correlated random variables, such as the time series of gene expression for a set of genes. Node labels in a DBN are typically indexed by the corresponding time point. In Figure 3.4, X_1, X_2, X_3 represent the random variable X at times $t = 1, 2, 3$, respectively. Here, because X_{t+1} is conditionally independent of X_{t-1} given X_t, the dynamical process is said to satisfy the *first-order Markov property* (Box 3.1). A concrete example of a DBN like the one in Figure 3.4 is a gene regulatory network, in which one can imagine $X_1, X_2, X_3, \ldots, Y_1, Y_2, Y_3, \ldots$ and $Z_1, Z_2, Z_3, \ldots$ as representing the expression levels of genes X, Y,

---

[2]Parent nodes of a node v are nodes connected to v via edges that are directed toward v.

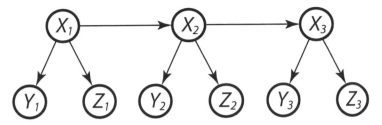

FIGURE 3.4: A dynamic Bayesian network representing three arbitrary random variables, X, Y, Z, either discrete or continuous, whose joint probability distribution changes with time. The network shows that Y and Z are conditionally dependent on the value of X at the same time point, whereas the distribution of X at a particular time point is conditionally dependent on its value at the previous time point. However, the distribution of Y at a particular time point is independent of its distribution at previous time points given the value of X at the same time point. Thus, for example, the figure implies that the joint distribution of $X_1, Y_1, Z_1, X_2, Y_2, Z_2$ is given by $P(X_1, Y_1, Z_1, X_2, Y_2, Z_2) = P(Y_1, Z_1|X_1)P(Y_2, Z_2|X_2)P(X_2|X_1)P(X_1)$.

and Z at different time points, and in which X regulates both Y and Z.

A DBN is completely specified by its architecture, that is, its connectivity diagram, and by the conditional probability distribution of each node given its parents. DBNs are capable of incorporating all the generalizations of Boolean networks that we have considered so far; they incorporate temporal data (by definition), noise in the form of conditional probabilities, and multiple expression states in the form of the set of values each node can take. There are also a number of choices for the scoring function in a DBN that affect the way network structure is inferred from gene expression data. Two of these choices are briefly discussed below.

3.4.1 Scoring and Network Inference with DBNs

To infer a DBN from expression data, one may imagine proceeding in a manner similar to that of the REVEAL method: find all singlets that explain the expression state of a gene G of interest, then find all doublets, triplets, and so on. These explanatory genes would then be the parents or inferred regulators of G. An important point of departure from the Boolean network is that the inferred explanatory relationship need not be a deterministic one. It is therefore not necessary to demand that the mutual information between G and its parent(s) in the

---

Box 3.1 Markov processes.

A stochastic process is a random process; that is, it is the time evolution of one or more random variables. A special type of stochastic process is the *Markov process* or Markov chain, in which the probability distribution of the random variable(s) under consideration at any given time point depends only on the probability distribution of the same random variable(s) at a finite number of previous time points. For simplicity, let the stochastic process consist of a single random variable X_t that is a function of time. Let us also assume that time progresses in discrete steps. We say that X_t is a first-order Markov process if the distribution of X_t depends only on the distribution of X_{t-1} and is conditionally independent of $X_{t-2}, X_{t-3}, \ldots$ given X_{t-1}. That is,

$$P(X_t \mid X_{t-1}, X_{t-2}, X_{t-3}, \ldots) = P(X_t \mid X_{t-1}).$$

Similarly, X_t is a second-order Markov process if its distribution depends only on the joint distribution of X_{t-1} and X_{t-2} but not on $X_{t-3}, X_{t-4}, \ldots$. In the Bayesian network picture, the process $X_1, X_2, X_3 \ldots$ shown in Figure 3.4 is a first-order Markov process because each X_t is conditionally dependent only on the previous X_{t-1}. If the Markov property is a good approximation for some random process, this constitutes a substantial simplification of the dynamics, as Markov processes are effectively memoryless beyond the previous time step. Many models of the dynamics of gene expression assume that the expression profile of a set of genes at any given time only depends on the expression profile at the previous time.

DBN be maximal, only that it be "large enough." How large is "large enough"? To figure this out, the significance of the magnitude of mutual information should be assessed using methods of statistical hypothesis testing. As discussed earlier in this chapter, the mutual information as estimated from finite data samples follows a χ^2 distribution when the variables in question are independent. Thus, a χ^2 test can be used to decide whether a certain amount of mutual information is statistically significant (Box 3.2).

It is customary to penalize the score S that is proportional to mutual information when the number of parameters inferred is large. This penalty is conceptually a consequence of the Occam's razor principle that favors simpler models over more complex ones. This principle is realized naturally in Bayesian statistics (a full discussion of this matter is beyond the scope of this book). The relevant score function that implements such a penalty is the BIC (Bayesian Information Criterion)

Box 3.2 The χ^2 statistical test.

Hypothesis testing in statistics is commonly carried out by assuming that the data are either in agreement with a *null* or an *alternative* hypothesis, in which the null hypothesis posits that the data are distributed according to some standard probability distribution. If the data have low probability under this standard distribution, we say that the null hypothesis has been rejected. In the conventional framework of hypothesis testing, one evaluates some function of the data (this function is called the *test statistic*), and gives the benefit of doubt to the null hypothesis by computing (assuming the null hypothesis) the probability of values more extreme in the direction of the alternative hypothesis than the observed value of the test statistic. This probability is the so-called *P-value*, and we reject the null hypothesis if the *P*-value is lower than some accepted threshold.

An example is provided by the χ^2 test for mutual information. As discussed in the text, the mutual information (which is the test statistic of interest) as estimated from data follows a χ^2 distribution when the underlying variables are independent. The probability density function of the χ^2 distribution is given by

$$f(x) = \frac{x^{\nu/2-1}e^{-x/2}}{2^{\nu/2}\Gamma(\nu/2)}, \quad x \geq 0,$$

where ν is an arbitrary parameter, called the number of *degrees of freedom* of the χ^2 distribution. Under the null hypothesis of independence, the mutual information is therefore distributed according to the above distribution. Under the alternative hypothesis of lack of independence, the mutual information would be expected to be higher. Thus, if the numerical value of the mutual information as estimated from data is S, the corresponding P-value is given by

$$Pr[x \geq S] = \int_S^\infty dx\, f(x) = 1 - F(S),$$

where $F(x)$ is the *cumulative distribution function* of the χ^2 distribution. The statistical test consists of rejecting the null hypothesis of independence if the *P*-value is very low, as a low *P* value means that the estimated mutual information has low probability under the null hypothesis.

The χ^2 test is not just used for testing the statistical significance of mutual information. More generally, it is a standard test for assessing the independence of two variables and appropriate in a large class of situations where the test statistic can be represented as a likelihood ratio. In its most basic form, it is a standard test of the variance of a normally distributed population.

score,

$$S_{BIC} = S - \frac{K}{2} \log N, \tag{3.14}$$

where K is the number of free parameters in the model and N is the number of data samples used to estimate the mutual information. The basic REVEAL method can be applied to find statistically significant values of S_{BIC} as one considers every set of putative parents of G. These statistically significant assignments can then be collected together to construct an inferred DBN that represents the regulatory network for the set of genes being studied (Friedman et al., 1998; Murphy and Mian, 1999).

An important issue with learning DBN models—one that also arises in simpler Boolean network models—is that of finding ways to limit the size of the search space of parent explanatory genes for a gene of interest. Many methods to do this are *heuristic* methods; that is, such methods do not guarantee that all explanatory genes can be found. One possibility is to examine only singlets and pairs of potentially explanatory genes for a gene of interest. All singlets and pairs that have statistically significant mutual information with the gene of interest can then be combined to yield the full set of parents of the gene of interest. Another heuristic method consists of starting with an initial guess for the regulatory network structure and then randomly carrying out local moves such as edge deletion, edge addition, or edge reversal that change the network. After each local move, the BIC score is recomputed and the move is accepted if the recomputed BIC score is higher than the previous one. This procedure leads to a network corresponding to a local maximum of the BIC score. Multiple initial guesses can then be combined to infer a global optimum.

3.4.2   Extensions, Variants, and Applications of DBNs

We now discuss further issues related to the application of the DBN methodology for accurate inference of regulatory networks. One of the problems with using expression data to infer regulation is that expression datasets are often incomplete; expression levels of important regulatory genes are sometimes missing because of experimental limitations. The situation is confounded by the fact that genes that are important are not known in advance. Hence, it appears that the standard DBN methodology would require knowledge of the expression levels of *all* genes that may play a role in the regulatory pathway under consideration. Fortunately, however, this knowledge is not always necessary;

the set of genes whose expression levels are missing may often be collectively characterized in terms of a small number of *hidden variables* or *hidden factors* that may, in principle, affect the expression states of genes whose expression levels are known. The problem of DBN inference then extends to simultaneously inferring both network structure and the states of these hidden variables.

The classic method of incorporating hidden variables in DBNs and inferring the underlying probabilities involves the Expectation-Maximization (EM) algorithm (Dempster et al., 1977; Lauritzen, 1995; Friedman, 1997; Friedman et al., 1998; Murphy and Mian, 1999), which is briefly summarized in Box 3.3. Using general results for EM algorithms, it can be shown that this iterative method for updating the parameters of a DBN always leads to an increase in the BIC score. The method is therefore guaranteed to converge to a DBN corresponding to a local maximum of the BIC score.

An EM-based approach was systematically used to model the gene regulatory networks that play an important role in T-cell activation (see Box 3.4 for a brief description of T-cells and the related B-cells) using expression data for about 100 human genes across ten time points (Rangel et al., 2004; Beal et al., 2005). The structure of the DBN model was constrained by assuming the following linear *state space model* for the dynamics of expression levels [see also Perrin et al. (2003)]:

$$\mathbf{g}_t = \mathbf{C}\mathbf{h}_t + \mathbf{D}\mathbf{g}_{t-1} + \mathbf{v}_t,$$
$$\mathbf{h}_t = \mathbf{A}\mathbf{h}_{t-1} + \mathbf{B}\mathbf{g}_{t-1} + \mathbf{w}_t, \tag{3.15}$$

where $\mathbf{g}_t$ is a vector of expression levels of all observed genes at time t, $\mathbf{h}_t$ represents a vector of hidden variables at time t, the components of $\mathbf{v}_t$ and $\mathbf{w}_t$ represent independent noise variables modeled as independent zero mean Gaussian (or Normally distributed) processes uncorrelated across time (this type of noise is termed *white Gaussian noise*), and $\mathbf{A}, \mathbf{B}, \mathbf{C}$, and $\mathbf{D}$ are matrices that characterize the linear relationships between the variables. These matrices are the parameters to be inferred. Note that within the DBN framework, all conditional probabilities are Normally distributed (because the noise variables are Normally distributed) and the means are linearly related according to the above equations. Furthermore, by substituting the second equation into the first, it is found that the effective matrix that captures the influence of $\mathbf{g}_{t-1}$ on $\mathbf{g}_t$ is $\mathbf{C}\mathbf{B} + \mathbf{D}$. Based on an entry in this matrix

---

Box 3.3 The Expectation-Maximization (EM) algorithm.

The Expectation-Maximization, or EM, algorithm (Dempster et al., 1977; Lauritzen, 1995; Friedman, 1997) is one of the most widely used methods to find parameters that maximize the likelihood function. In a typical approach to the problem, one must infer unknown parameters θ that describe the data in some way. The data themselves consist of observed variables $\mathbf{x}$ and "hidden" variables $\mathbf{y}$. For concreteness, the reader can imagine θ as signifying unknown regulatory relationships among genes, $\mathbf{x}$ as observed expression levels, and $\mathbf{y}$ as unobserved expression levels. The method iteratively improves upon an initial guess $\theta^{(0)}$ for the parameter values, as follows.

The likelihood function is given by $L(\theta) = P(\mathbf{x}, \mathbf{y} \mid \theta)$. It is not possible to directly maximize $L(\theta)$ (and thus estimate θ) because the values of the hidden variables $\mathbf{y}$ are unknown. Instead, let us suppose that the current best guess for θ is $\theta^{(i)}$. Using this, one can compute the expected, or mean, values of the hidden variables *given* the values of the observed variables and the parameters $\theta^{(i)}$. This is the "E" step in the EM algorithm; it results in an estimate of the values of the hidden variables. Let us call this estimate $\mathbf{y}^{(i)}$. The next, "M" step, consists of maximizing the likelihood function using the observed values $\mathbf{x}$ and the estimated values $\mathbf{y}^{(i)}$. The parameter value(s) at which this maximum is attained is the new estimate for θ, which we call $\theta^{(i+1)}$. The same procedure is then repeated until reasonable convergence of the parameter values is attained.

It can be shown that the EM procedure always increases the likelihood function. Also, even in cases where no hidden variables exist, it can be mathematically convenient to describe some of the unknown parameters as hidden variables if that redefinition leads to a likelihood function that can be maximized in a straightforward manner.

---

being significantly positive or significantly negative, that entry can be interpreted as up- or down-regulation, respectively.

Figure 3.5 shows one of the inferred sub-networks implicated in T-cell activation, as found by Rangel et al. (2004). The gene *fyb* turns out to be the central gene in the T-cell activation network predicted by Rangel et al., as it is predicted to regulate the largest number of genes. The genes shown in the sub-network are all directly related to the immune response. The *fyb* gene is known to play an important role in the signaling mechanism associated with the T-cell receptor molecule on the surface of the T-cell. The three interleukin receptor genes *IL2Rγ*,

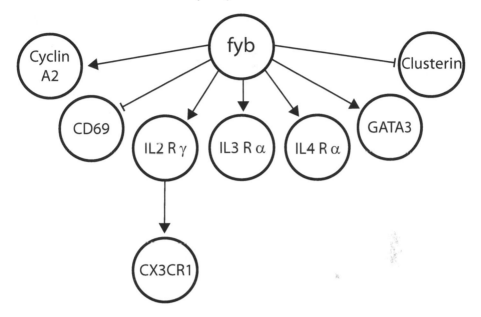

FIGURE 3.5: One of the inferred sub-networks implicated in T-cell activation (Rangel et al., 2004). This network contains genes that are downstream of gene *fyb*. Positive regulations are represented by arrowheads; negative regulations are represented by bars.

IL3Rα, and *IL4Rα* encode for the IL-4 receptor protein whose levels are known to be suppressed when *fyb* expression is diminished, in agreement with the predicted sub-network. The prediction that the chemokine receptor CXC3R1 is up-regulated by IL-2 receptor gamma is also consistent with data from other cell types. Further details on the plausibility of some of the other predictions and on the biological hypotheses suggested by them are given in the work of Rangel et al. (2004).

One of the problems with the DBN methodology is that it can be significantly expensive computationally and does not scale well with the number of nodes (here, genes) in the DBN. It is therefore often necessary, when trying to infer networks from large expression datasets, to reduce the effective number of genes in the problem. This can be done in a number of ways, the simplest among them involving pre-processing the data to remove genes that have very low expression levels throughout the time course of the experiment, or genes that have very low variation in expression level across the time course, or

Box 3.4 T-cells and B-cells.

T- and B-cells are special types of white blood cells that play an important role in the immune response system. T-cells derive their name from their origin in the thymus gland. Depending on their type, they have a number of functions, including that of aiding in the maturation of B-cells and the destruction of virally infected cells and tumor cells. Their activation involves a large number of external signals and the expression of many genes in temporal sequence. The B-cells, produced in the bone marrow of most mammals, mature in response to signals from T-cells and principally become plasma cells or memory B-cells. Plasma cells are responsible for the production of antibodies against bacterial proteins or viruses (antigens). Memory B-cells retain genetic rearrangements that allow for the production of specific antibodies against specific antigens (i.e., they retain "memory" of a previously encountered antigen).

by clustering together genes that have similar expression profiles into a single "meta-gene"—this last operation tacitly assumes that all genes represented by the meta-gene will serve as common regulators of other genes and/or be simultaneously regulated by other genes.

A novel approach to reduce the number of putative regulators as well as estimate transcriptional time lags in regulation more accurately was suggested by Zou and Conzen (2005). Their method consists of first identifying all potential regulators of a gene G of interest as those genes whose expression level first changes before or concurrently with the expression level of G. Next, potential regulators are clustered according to the difference in time of first change in expression level between the regulator and G. This time difference is taken to be the transcriptional time lag between the regulator and G. Subsets of genes belonging to a common cluster of potential regulators (i.e., potential regulators that have the same transcriptional time lag with respect to G) are then examined for high BIC scores, with the best subsets from every cluster pooled together and inferred as true regulators of G. The process is repeated for every gene G.

The method of Zou and Conzen (2005) implicitly has a pre-processing step (i.e., filtering by transcriptional time lag) that reduces the number of potential regulators of G. Furthermore, only subsets within clusters are examined for the magnitude of their BIC score with G, thus greatly increasing the efficiency of the method. Zou and Conzen tested this method on a yeast cell-cycle time course expression dataset that included many previously established regulatory genes. They reported substantial increases in accuracy and substantially lower computational time using their method relative to the standard DBN approach.

Other than the applications mentioned above, the reader should be aware that there is a vast literature on the use of DBNs for the inference of regulatory networks. This includes a systematic analysis of the accuracy of DBNs when applied to microarray gene expression data (Husmeier, 2003), an adaptation of the DBN methodology for addressing data from perturbation experiments (Dojer et al., 2006), and an application to nonlinear, nonparametric models (Kim et al., 2004).

3.5   Predicting Protein-Protein Interactions

As we have discussed, no single high-throughput experimental method for discovering protein-protein interactions is currently reliable. Also, while high-throughput experiments often do not agree with each other, the interactions that are consistently reported by different experiments are usually correct. This emphasizes the need for various ways of combining high-throughput results in order to make sense of interaction data, a topic we examine in greater detail in Chapter 5. The inefficiency and lack of reliability of experimental approaches also opens the door for computational approaches that can realistically aim to predict protein-protein interactions at least as reliably as high-throughput experiments.

A number of factors influence the propensity of two proteins to partake in a physical interaction. In keeping with convention, we call these factors *features* that correlate with the presence of physical interaction. Many of these features are derived from the sequence, structure, and evolutionary history of proteins. Some of them are derived based on the spatio-temporal context in which proteins occur in the cell. For example, interaction between two proteins is possible only if they are both present in the cell in comparable quantities in the same cellular compartment at the same time (i.e., they are co-localized and co-expressed). Interacting proteins are also likely to have correlated patterns of mutation at the interaction site, as any point mutation in one protein is likely to be accompanied by a compensatory mutation in the partner protein in order to stabilize the interaction.

Here, we summarize (by no means exhaustively!) some of the features that have been used to predict protein-protein interactions *in silico*. Detailed reviews on this subject include those of Skrabanek et al. (2008) and Liu et al. (2008). We later describe how these and other simpler and weakly predictive features may be integrated to improve prediction performance and render the prediction task more robust to errors in individual feature sets.

3.5.1   Structural Features

The earliest methods for predicting protein-protein interaction utilized the tertiary structure of the proteins involved, and hence could be applied only to the small proportion of proteins whose three-dimensional structure is known. There are two aspects to structure-based prediction of protein-protein interaction: prediction of interaction sites within individual proteins, and prediction of the propensity for two given proteins to actually interact. The first aspect is nonspecific, as it refers only to single proteins.

As a case in point, it has been found that sites of interaction tend to be predominantly populated by hydrophobic residues that lie on the protein surface and are therefore exposed to solvent. The occurrence of hydrophobic residues on part of the surface creates a tendency for that part of the surface to exclude water and thus partake in a binding interaction with a similar surface on another protein. The excluded water molecules form a cage-like structure surrounding the interaction surface, thus contributing to the overall stability of the bound protein complex. This nonspecific tendency for interaction can, in fact, be used to predict interaction networks whose statistical properties appear to agree well with those obtained from high-throughput experiments. This aspect is discussed in greater detail in Chapter 7.

Other parameters that are used to distinguish residues at interaction sites from other residues include geometrical features such as planarity and protrusion of a "surface patch" around the residue in question (Jones and Thornton, 1997a). These geometrical features can be combined with hydrophobicity, solvation potential, accessible surface area, and residue interface propensity (computed from the statistics of residues that are known to lie along protein interaction interfaces) to enable reasonably accurate prediction of interaction sites (Jones and Thornton, 1997b). Figure 3.6 gives an example of the assessment of interaction propensity for surface patches on the tryrosyl-tRNA synthetase protein, an important enzyme involved in the addition of tyro-

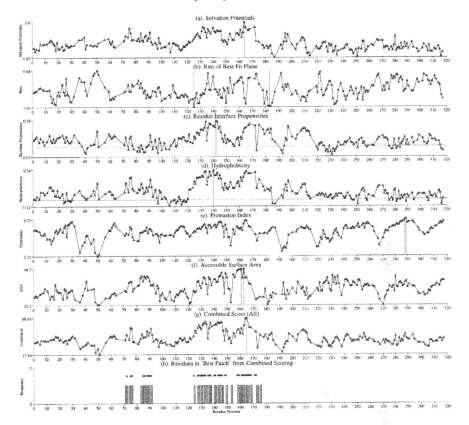

FIGURE 3.6: "Patch profiles" of tyrosyl tRNA synthetase (Protein Data Bank ID: 2ts1) using six parameters that distinguish interacting surface patches from noninteracting ones. In each profile (a)–(f), residue numbers are indicated on the x-axis and the value of the parameter on the y-axis. The six profiles are integrated into a Combined Score profile (g) that yields predictions for residues involved in protein-protein interaction (h). Residues on the known interface are marked by an asterisk ($*$) above the vertical bars in panel (h). Reproduced with permission from Jones and Thornton (1997a).

sine to the protein chain during protein synthesis.

While the features mentioned above are useful for determining the generic propensity of a part of a protein to be involved in protein-protein interactions, it is often of interest to predict the propensity of two *specific* proteins to interact. Structural methods to predict the propensity of interaction between two specific proteins are often based on complementarity between the surfaces in question. For example,

---

Box 3.5 Molecular docking.

Molecular docking is a computational method that predicts the relative orientation of one molecule when bound to another molecule to form a complex. Docking is most commonly carried out in the initial stages of rational drug design, where a candidate small-molecule (i.e., a *ligand*) drug is "docked" on a target protein molecule. It is also used in the elucidation of protein-protein interactions, where one protein is docked to another. One of the most important outcomes of a docking analysis is a measure of the binding affinity between the two molecules studied. This affinity is called the *docking score*. While docking scores often do not approximate the free energy of binding very well, one of the principal advantages of a docking simulation is that it is computationally fast and can be used to quickly screen large numbers of candidate ligands or proteins and therefore generate a short list of molecules that are expected to bind stably with the protein of interest.

---

shape complementarity is used in docking studies (Box 3.5) that find the best fit of two interacting proteins (Lawrence and Colman, 2003). A public resource that combines the features used to predict interaction sites with shape complementarity analysis in order to predict protein-protein interactions is the *Protein-Protein Interface Analysis Server* (http://www.bioinformatics.sussex.ac.uk/protorp/).

At the time of writing, it appears that structural properties are poised to play a central role in the prediction of protein-protein interactions. While this development may seem obvious given that protein-protein interactions *in vivo* are almost entirely determined by structural features, progress in this area has to date proved difficult because of the lack of availability of high-resolution structures as well as the lack of accurate algorithms and computational power to determine whether two proteins of known (high-resolution) structure will actually interact. However, the growth of structural databases such as the Protein Data Bank (PDB, http://www.rcsb.org), in concert with the growth of structural genomics initiatives that aim to infer the structures of proteins of unknown structure by homology to proteins of known structure, and the attendant rise in computational capabilities for studies of protein-protein docking, has made the prospect of protein-protein interaction prediction using structural information very attractive.

As a first step in this direction, Wass et al. (2011) carried out a high-throughput docking study in which 56 proteins known to participate in complexes were docked against a nonredundant set of 922 monomeric proteins, some of which are known to interact with the proteins in the first set, but most of which do not do so. The aim of the study was to assess whether docking scores can reliably discriminate between true and false interactors. The three-dimensional structure of all proteins in the study was known to high resolution. The study required tremendous computing power because the docking score in each case depends on the "pose" (relative orientation in three-dimensional space) of the two proteins being docked. A large number of poses had to be generated for each protein pair, leading to a score distribution for each protein pair. Their results were encouraging in that they showed that the score distribution of known interactors is often distinguishable from that of the background, and furthermore the score distribution of known interactors was also distinguishable from other proteins of very similar structure (Figure 3.7).

3.5.2 Gene Neighborhood Analysis

The central hypothesis behind prediction of protein interaction from gene neighborhood is that genes corresponding to proteins which physically interact will be in close proximity to each other on the genome. Indeed, operons in bacteria, which are simultaneously transcribed sets of genes located in a single cluster on the genome, code for proteins that are functionally associated and therefore likely to interact. Although operons are rare in eukaryotes, genes that are repeatedly observed in close proximity (within about 500 base-pairs of each other) are potential candidates for functional association or protein-protein interaction.

The disadvantage of this method of inferring interactions is that it does not distinguish between functionally associated proteins and interacting proteins. Furthermore, the method is generally more successful on prokaryotic than eukaryotic proteins, because co-regulation in eukaryotes is more complex and less influenced by gene proximity. In bacteria and archaea, the method has been successfully used to identify new members of metabolic pathways (Dandekar et al., 1998). The method becomes more powerful as larger numbers of genomes are available for neighborhood analysis, because interacting proteins in one organism can be predicted by their gene proximity in combination with evidence of interactions among homologous proteins in other organisms (see the discussion on interologs below).

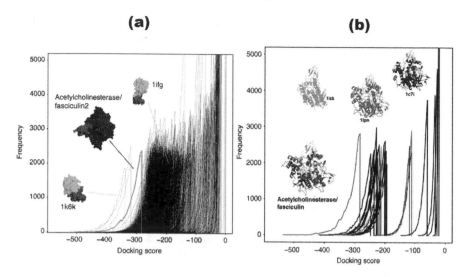

FIGURE 3.7: Docking score distribution of the acetylcholinesterase-fasciculin2 complex [curve indicated in panel (a); the leftmost curve in panel (b)] compared with docking score distributions of (a) a background set of putative complexes, and (b) putative complexes of fasciculin2 docked with members of the α/β-hydrolases structural superfamily, to which acetylcholinesterase belongs. In (a), two additional selected background models are shown. In (b), the docking score distributions for the three most structurally similar proteins to acetylcholinesterase are shown with their structures. While the type of separation of scores between true and putative complexes is not typical, this example conveys the potential of high-throughput docking to discover protein-protein interactions. Reproduced with permission from Wass et al. (2011).

.

    The STRING (Search Tool for the Retrieval of INteracting Genes/proteins) database, available at http://string.embl.de, combines neighborhood analysis with other sources to predict protein-protein interactions.

3.5.3 Phylogenetic Profiles

Like gene neighborhood analysis, similarity of phylogenetic profiles (Box 3.6 and Figure 3.8) is primarily indicative of functional similarity or association. Similarity between two phylogenetic profiles, also termed *co-evolution*, indicates functional association or physical interaction between the proteins in question. However, this simple approach

Box 3.6 Orthology and phylogenetic profiling.

Genes in different organisms are said to be *orthologous* if they can be evolutionarily traced back to the same ancestral gene. This ancestral gene must predate the time of speciation (i.e., separation of one species into two) between the two organisms in question. Following speciation, identical genes in the two organisms are subject to independent mutation events over the course of evolution, thus leading to *divergence* between their DNA sequences. Because we cannot directly observe ancestral genomes, orthologous relationships among genes are inferred by the similarity of their DNA or protein sequences. Thus, when we refer to the same gene in different organisms, we really mean orthologous genes, or *orthologs*. Sometimes orthologs are also referred to as *homologs*, although strictly speaking, homology encompasses any form of similarity, not just that arising from evolutionarily common ancestry.

The study of evolutionary relationships between organisms is called *phylogeny*. In its simplest form, a *phylogenetic profile* of a gene across a set of organisms is mathematically a binary vector of 0s and 1s that signifies the occurrence (1) or lack of occurrence (0) of the (orthologous) gene in each organism of interest. A *similarity score* between the phylogenetic profiles of two genes can be defined as the proportion of entries in the two vectors that have the same value.

is sensitive to the number and distribution of genomes used to build the profile. Genes that co-occur in a small number of closely related species are less likely to code for interacting proteins than genes that co-occur in a large number of distantly related species.

The co-evolution method for inferring interactions is also sensitive to the manner in which orthologous sequences are detected in the different organisms, and is problematic in principle because co-evolution could simply occur in a nonspecific manner. For example, any two proteins in the same organism will share a background level of co-evolution that they inherit from the speciation of the whole organism itself. Thus, co-evolution need not be specifically associated with protein-protein interaction.

An elegant and powerful way of removing false positives associated with nonspecific co-evolution was suggested by Juan et al. (2008), and is depicted in Figure 3.8. Briefly, after computing the similarity score between phylogenetic profiles of two proteins A and B, the extent of correlation (as computed using, say, the Pearson correlation coefficient) between the similarity scores of A with all other proteins in the organism and the similarity scores of B with all other proteins in the organism is found. This correlation measures the extent of similarity between the *co-evolutionary networks* of A and B. If such a correlation is high, that strongly suggests an association between A and B over and above their isolated co-evolution. As a further filtering step, Juan et al. (2008) proposed to eliminate nonspecific correlations between the co-evolutionary networks by asking whether an observed correlation between A and B can be explained by their mutual correlations with the co-evolutionary network of a third protein. This last step in the analysis is carried out using partial correlation coefficients (see Box 2.10). Only protein pairs whose similarity of co-evolutionary networks cannot be explained by other intermediate correlations are considered candidate interacting proteins. This method impressively provides a degree of accuracy and coverage of protein-protein interactions comparable to that of experimental techniques.

3.5.4 Gene Fusion

Gene fusion is a common evolutionary event whereby two separate parent genes in an ancestral organism are fused in an extant organism into a single multifunctional gene that typically codes for a multidomain protein. Two separate proteins in a *query* organism that show significant similarity to different regions of a single, fused protein in a *reference* organism are likely candidates for functional or physical association in the query organism, as they occur in fused form in the reference organism. In a gene fusion analysis, it is, however, crucial that the two proteins in the query organism are not significantly similar to each other. If they are, they will naturally show significant similarity to the same protein in the reference organism even if no gene fusion occurred in the reference protein. Thus, if the two proteins in the query organism are significantly similar to each other, they are very likely to correspond to *paralogs*, or duplicate genes/proteins, rather than interacting proteins.

A possible source of error in the prediction of interacting proteins from gene fusion events has to do with fusion by recombination. There are situations where two functionally unrelated protein domains become fused by chance in the reference organism due to a recombination event (leading to a two-domain protein of novel function), in which case

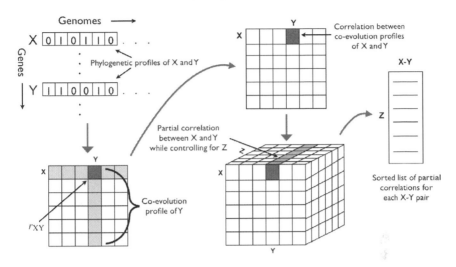

FIGURE 3.8: Inference of protein-protein interaction by co-evolutionary analysis. Correlations between phylogenetic profiles of all genes in a genome are collected to generate a co-evolution profile for each gene. Next, correlations between co-evolution profiles of genes are found. These indicate the extent to which two genes share similarity of evolutionary histories with other genes. Finally, partial correlations are used to examine the extent to which correlated co-evolution can be explained by a third other gene. Gene pairs whose correlation in co-evolution is not significantly explicable by the co-evolutionary profile of a third gene are considered candidates for protein-protein interaction. Adapted with permission from Juan et al. (2008).

there is no reason why the homologous domains in the query organism should function in tandem or interact with each other.

The gene fusion method has been successfully applied to reveal a number of novel protein-protein interactions in the fruit fly *Drosophila melanogaster* (Iliopoulos et al., 2003). The FusionDB database (http://www.igs.cnrs-mrs.fr/FusionDB/) contains, at the time of writing, the results of all annotated genes in 51 query genomes checked for fusion events in 89 fully sequenced bacterial and archaean genomes.

3.5.5 History of Correlated Mutations

It has been observed that a mutation in the sequence of one protein in a pair of interacting proteins is often accompanied by a compensatory mutation in its interacting partner protein. This phenomenon of occurrence of correlated mutations at interaction sites is also observed in single-stranded RNA, where correlated mutations occurring along a single RNA sequence may be used to successfully predict base-paired nucleotides, and hence the RNA secondary structure.

The correlated mutation method as applied to a pair of proteins (Jothi et al., 2006) identifies an aligned family of homologous proteins for each member of the pair. If there are positions in each alignment that show significant correlation across aligned residues, the two proteins are considered candidates for interaction. Moreover, the positions of the aligned residues identify the interaction surface. Typically, the method is quite sensitive to the quality and number of alignments. This approach has been termed the *in silico two-hybrid* method for finding protein-protein interactions (Pazos and Valencia, 2002).

Figure 3.9 gives a schematic overview of some of the methods described above. Note that these methods are, in a sense, *ab initio* methods for deciphering protein-protein interactions, as they do not rely on the knowledge of other protein-protein interactions.

3.5.6   Existence of "Interologs"

One of the strongest predictors of protein-protein interactions in one organism is the existence of a pair of orthologous proteins that is known to interact in another organism. Such conserved pairs of interacting proteins (Figure 3.10) are termed "interologs" (Walhout et al., 2000; Matthews et al., 2001).

Matthews et al. (2001) carried out a systematic study of the extent to which searches for interologs may be used to identify protein interaction networks in the organism of interest. They used two two-hybrid *Saccharomyces cerevisiae* protein interaction maps to predict interactions in the nematode worm *Caenorhabditis elegans*, and validated these predictions via two-hybrid experiments. In general, the interolog method can be used to reliably infer a protein-protein interaction between proteins A and B based on a known interaction between proteins A' and B' in a different organism when the geometric mean of the sequence identities of A-A' and B-B' is at least 80% (Yu et al., 2004).

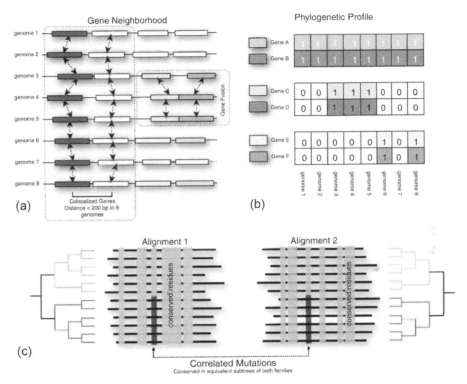

FIGURE 3.9: *Ab initio* approaches to prediction of protein-protein interaction based on genomic information. (a) A pair of genes (shown in dark gray and white) in close physical proximity in eight different genomes is suggestive of association or interaction between the proteins. Gene fusion between two genes in genomes 4 and 5 is similarly suggestive of association between the corresponding proteins in genome 3. (b) Phylogenetic profiles of six genes, with three interacting protein pairs (A-B, C-D, E-F) inferred because of similarity in these profiles. (c) Two protein families with conserved regions highlighted in light gray. Correlated mutations (highlighted in dark gray) indicate regions of mediating interactions between proteins from each family. Reproduced with permission from Skrabanek et al. (2008).

3.6 Data Integration to Improve Prediction Performance

The features discussed above carry relatively strong signals for protein-protein interaction, and have been used individually to predict interactions. However, there are a number of weak signals or weak "correlates" of protein-protein interaction that could perhaps be combined to yield

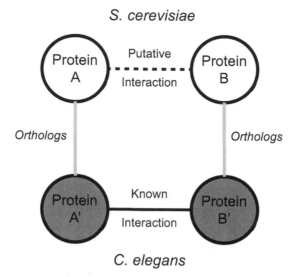

FIGURE 3.10: An illustration of the interolog concept. Proteins A' and B' in the worm *Caenorhabditis elegans* are known to interact. Proteins A and B in the yeast *Saccharomyces cerevisiae* are orthologs of A' and B', respectively. An interaction between A and B can be inferred based on the known interaction between A' and B'. The A'-B' pair is called an *interolog* of the A-B pair.

a combined predictor that is powerful. These signals are typically positive for interacting protein pairs but do not by themselves provide convincing evidence that the two proteins in question actually interact.

Examples of such weak signals include subcellular co-localization and co-expression (as mentioned earlier); essentiality (proteins in an interacting pair are typically both essential or both nonessential, as a deletion mutant of either one of the two should produce roughly the same phenotype); biological function (interacting proteins often participate in the same biological process); and mutual clustering (interacting proteins often tend to share their interacting partners; see Section 4.4). These weak correlates are individually far more useful for *validating* experimentally derived protein-protein interactions rather than for actually *predicting* protein-protein interactions. Here we focus on how such a set of weak correlates may be intelligently integrated to produce a powerful predictor for protein-protein interactions.

A host of increasingly sophisticated methods for data integration currently exist, and continue to be a fruitful topic of current research. All of these methods involve two procedures: a procedure for quantitatively representing or encoding each correlate, and a procedure for combining the numerical representations into a predictive measure. Co-expression, for example, could be encoded in terms of a number of different correlation or expression profile similarity measures. Similarly, some data integration methods work best with discrete representations of the underlying data. In these cases, continuous variables must be discretized or "binned" before they are fed into the appropriate data integration method.

3.6.1 Bayesian Networks and Feature Integration

One of the seminal examples of an integrated method to predict protein-protein interactions consists of a Bayesian network method devised by Jansen et al. (2003). We have already seen how dynamic Bayesian networks may be used to predict regulatory interactions from gene expression data. The method of Jansen and co-workers is significantly simpler yet powerful, and essentially uses a Bayesian representation for the probability of a protein-protein interaction given multiple sources of weak evidence.

Consider N features, $f_1, f_2, \ldots, f_N$, each of which by itself is weakly predictive for protein-protein interaction. If the numerical value of each feature is known for a pair of proteins, the Bayesian network method involves calculation of the *likelihood ratio* $L(f_1, \ldots, f_N)$, defined as

$$L(f_1, \ldots, f_N) = \frac{P(f_1, \ldots, f_N \mid \text{pos})}{P(f_1, \ldots, f_N \mid \text{neg})}, \qquad (3.16)$$

where $P(f_1, \ldots, f_N \mid \text{pos})$ is the probability of the observed feature values if the protein pair were interacting, or "positive," and similarly $P(f_1, \ldots, f_N \mid \text{neg})$ is the probability of the observed features if the protein pair were noninteracting, or "negative." A large likelihood ratio means that the evidence is more consistent with the presence of an interaction rather than the absence of it; conversely, small values of the likelihood ratio indicate that the data are consistent with the *absence* of interaction.

How do we compute the likelihood ratio, and how large should it be for a pair of proteins to be predicted as interacting? To answer the first part of the question, we consider a set of known, or "gold-standard" positive and negative (i.e., interacting and noninteracting) protein pairs, compute numerical values of the features for each pair, and then bin these numerical values separately for positive and negative pairs to generate, for each feature "bin", the two probabilities required to compute the likelihood ratio for that feature bin. This procedure yields a table of likelihood ratio values for each feature value bin. All that is required to then find the likelihood ratio for a pair of proteins is to compute the feature values for that pair, identify the correct bin it belongs to, and read off the likelihood ratio from the table constructed using gold-standard pairs.

The procedure for constructing likelihood ratios for gold-standard pairs is greatly simplified if each feature is *conditionally independent* of the others. Conditional independence means that information in each of the N features of a protein pair is independent of any other feature, given that the protein pair is either positive or negative. Mathematically, the assumption of conditional independence leads to factorization of the two probabilities in Equation (3.16), thus yielding $P(f_1, \ldots, f_N \mid \text{pos}) = \prod_{i=1}^{N} P(f_i \mid \text{pos})$ and $P(f_1, \ldots, f_N \mid \text{neg}) = \prod_{i=1}^{N} P(f_i \mid \text{neg})$. If the assumption of conditional independence is reasonable, then the Bayesian network is called a "naive" network, the method is called the *naive Bayes method*, and the likelihood ratio can be simplified to

$$L(f_1, \ldots, f_N) = \prod_{i=1}^{N} L(f_i) = \prod_{i=1}^{N} \frac{P(f_i \mid \text{pos})}{P(f_i \mid \text{neg})}, \qquad (3.17)$$

that is, the full likelihood ratio is a product of the likelihood ratios corresponding to each feature considered individually.

Jansen and co-workers used a "gold-standard" dataset of interacting and noninteracting proteins in conjunction with conditionally independent features such as essentiality, correlation of mRNA expression profiles, functional similarity according to the Munich Information center for Protein Sequences (MIPS), and similarity of biological process according to the Gene Ontology database. In addition, they also considered features that were not conditionally independent (the incorporation of these features mathematically leads to a slightly different reformulation of the problem): for example, evidence of protein-protein interaction from different high-throughput experiments. For constructing the gold-standard dataset, a protein pair was considered positive if it was part of the same protein complex according to MIPS (strictly speaking, the method of Jansen and co-workers only predicts if two proteins will be part of the same complex), and a protein pair was considered negative if the two members of the pair were found in two different subcellular compartments. A sample calculation of the likelihood ratio using just three features is illustrated in Box 3.7. This type of calculation can be extended to arbitrary numbers of features, and likelihood ratio values for a large number of feature values can be pre-computed.

For the protein pair on which prediction is to be carried out, the numerical values of each feature are computed, and the corresponding individual likelihood ratios are read from the pre-computed table. These ratios are then multiplied to yield the full likelihood ratio. If this full likelihood ratio exceeds a certain threshold, the pair is predicted to be interacting (or more precisely, part of the same protein complex), otherwise not.

This brings us to the second part of our original question, namely how to decide the value of the likelihood ratio threshold. To address this, we turn to the Bayesian interpretation of the likelihood ratio. First one defines the prior odds ratio,

$$O_{\text{prior}} = \frac{P(\text{pos})}{P(\text{neg})} = \frac{P(\text{pos})}{1 - P(\text{pos})}, \tag{3.18}$$

where $P(\text{pos})$ is the prior probability that a protein pair is interacting, or positive. This is found from estimates of the sparseness of the protein interaction network in the organism of interest. In the case of yeast, it is thought that there are roughly 30,000 pairs of proteins in the same complex out of 18 million total pairs, leading to $O_{\text{prior}} \simeq 1/600$.

Next, we define the posterior odds ratio,

$$O_{\text{post}} = \frac{P(\text{pos} \mid f_1, \ldots, f_N)}{P(\text{neg} \mid f_1, \ldots, f_N)}. \tag{3.19}$$

Using Bayes' rule, we therefore find that the likelihood ratio can be expressed as

$$L(f_1, \ldots, f_N) = \frac{O_{\text{post}}}{O_{\text{prior}}}. \tag{3.20}$$

Now, in order for a protein pair to be classified as positive given values of the features for that pair, it must be that $O_{\text{post}} > 1$. Using our estimate of O_{prior}, this translates to $L > 600$. Thus, a yeast protein pair is considered positive if its full likelihood ratio exceeds about 600. Figure 3.11 shows three protein complexes predicted by the method that were partially verified by TAP-tagging.

In addition to the Bayesian network data integration method outlined above, a number of other sophisticated methods exist, most of which are beyond the scope of this book. Logistic regression models to predict

---

Box 3.7 Application of likelihood ratio calculations to the prediction of protein-protein interactions: An example.

Consider the feature corresponding to the *essentiality* of two proteins and let us examine its predictive value for protein-protein interaction. There are three possible combinations: either both members of a protein pair are essential (corresponding to feature value EE), only one of them is essential (feature value EN), or both are nonessential (feature value NN). In the "gold-standard" interaction dataset constructed by Jansen and co-workers, there are 2,150 interactions in which both members are clearly annotated for essentiality, that is, each of the 2,150 interactions can be labeled as one of EE, EN, or NN. Similarly, there are 573,724 noninteractions that are clearly annotated for essentiality. Of the 2,150 annotated "positives," 1,114 are EE, 624 are EN, and 412 are NN. Thus, $P(EE \mid \text{pos}) = 1{,}114/2{,}150 = 0.518$, $P(EN \mid \text{pos}) = 624/2{,}150 = 0.290$ and $P(NN \mid \text{pos}) = 0.192$. Similarly, of the 573,724 annotated "negatives," 81,924 are EE, 285,487 are EN, and 206,313 are NN, yielding $P(EE \mid \text{neg}) = 0.143$, $P(EN \mid \text{neg}) = 0.498$, and $P(NN \mid \text{neg}) = 0.360$. The likelihood ratio for two essential proteins is therefore given according to Equation (3.17) by $L(EE) = P(EE \mid \text{pos})/P(EE \mid \text{neg}) = 0.518/0.143 = 3.6$. Similarly $L(NE) = 0.6$ and $L(NN) = 0.5$. Although the likelihood ratio for two essential proteins is larger than 1, the discussion below Equation (3.20) shows that we cannot predict a pair of yeast proteins to be interacting based on their essentiality alone because the likelihood ratio computed using just the essentiality feature does not exceed 600.

Now let us consider a second feature, namely the correlation r of mRNA expression levels between two proteins, and let us say that the two proteins have highly correlated expression if $r > 0.8$. Using expression data for proteins in the gold-standard interaction dataset, it is found that the proportion of interacting protein pairs having highly correlated expression is about 0.02 or 2%, while the proportion of noninteracting protein pairs having highly correlated expression is about 2.3×10^{-4} or 0.023%. Thus, the likelihood ratio for two proteins with $r > 0.8$ is $L(r > 0.8) = 0.02/0.00023 = 87$. Similarly, a third feature, the similarity of biological function among two proteins, can be analyzed to show that the likelihood ratio for two proteins having highly similar biological function is $L = 25.5$.

Note that the three features we have discussed are individually only weakly predictive of protein-protein interaction. The likelihood ratio based on any one of the three features cannot exceed 600. However, a pair of proteins in which both members are essential, have high expression correlation and high similarity of biological function has a likelihood ratio of $3.6 \times 87 \times 25.5 \simeq 7{,}987$. These proteins can therefore be predicted to be interacting proteins.

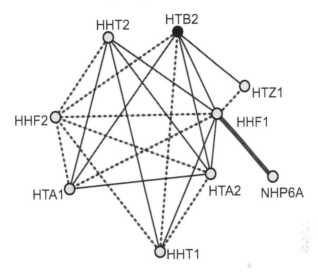

Nucleosome Proteins

FIGURE 3.11: A complex of nucleosome proteins in yeast as predicted by the Bayesian network method of Jansen et al. (2003) using only gene expression, gene function, and essentiality data. The complex contains proteins linked to a single central protein (black node). The single interaction verified by TAP-tagging alone is shown as a thick solid line, interactions previously known from high-throughput experiments (this knowledge was not used for the prediction) are shown as dashed lines, and overlaps between the two are shown as thin solid lines.

protein-protein interactions have been used by Lin et al. (2004) and Sharan et al. (2005). Random forest methods have been used by Lin et al. (2004) and Qi et al. (2006), who also compared the performance of random forest methods to support vector machines. A good survey of data integration methods for prediction of protein-protein interaction is the review article by Liu et al. (2008).

Finally we should also mention integrative methods to predict domain-domain interactions from protein-protein interactions. Protein domains are structural and functional subunits of proteins that are evolutionarily conserved. Distinct proteins may be composed of different combinations from the same set of domains. Therefore, multifunctional proteins are typically multidomain proteins. Also, protein-protein interactions are typically mediated by specific domain pairs. Identification of interacting pairs of domains is thus a powerful tool for predicting protein interactions among novel proteins that possess those domains.

Methods to identify interacting domain pairs include maximum likelihood estimation (Deng et al., 2002; Liu et al., 2005), Bayesian methods (Lee et al., 2006; Kim et al., 2007), domain pair exclusion analysis (Riley et al., 2005), and parsimony explanation (Guimarães et al., 2006).

Chapter 4

Metabolic Networks and Genetic Interactions

We first encountered metabolic networks in Chapter 1. Unlike the physical interaction networks discussed in the two previous chapters, metabolic networks are somewhat more abstract. From the discussion in Chapter 1, we know that metabolic pathways depict series of chemical reactions that modify chemical substances (metabolites) and in the process synthesize the basic building blocks of the cell, including nucleic acids, amino acids, and fatty acids. By breaking down metabolites, these reactions use the energy stored in chemical bonds. Thus each edge in a metabolic network is a meta-edge comprising at least three individual physical interactions: $enzyme_i$-$substrate$, $enzyme_i$-$product$, $product$-$enzyme_{i+1}$, where the $product$ is really the $substrate$ of the next enzyme in the pathway (Figure 4.1). The concentrations of the various entities that make up metabolic networks are governed by the chemical rate equations for the reactions these entities participate in, and the form of these rate equations is in turn determined by the connectivity diagram (topology) and stoichiometry of the components of the metabolic network. The edges of a metabolic network may be bi- or unidirectional depending on the particular rates and equilibrium constants of the individual reactions (Box 4.1).

Metabolic networks are special because they offer a direct testing ground for the connection between genotype and phenotype. At the genotypic level, a gene encodes a particular enzyme, and changes in gene expression therefore lead to changes in the concentrations of the corresponding enzyme products. If detailed reaction parameters for all of the reactions these enzymes catalyze are known, one can predict the corresponding changes in metabolite levels. These changes often lead to phenotypical changes in the organism of interest. For example, reduced expression of enzymes that characterize anabolic reactions will lead to low reaction rates for these reactions and therefore to slow cell growth. In a more complex example, consider a mutation in the gene for the en-

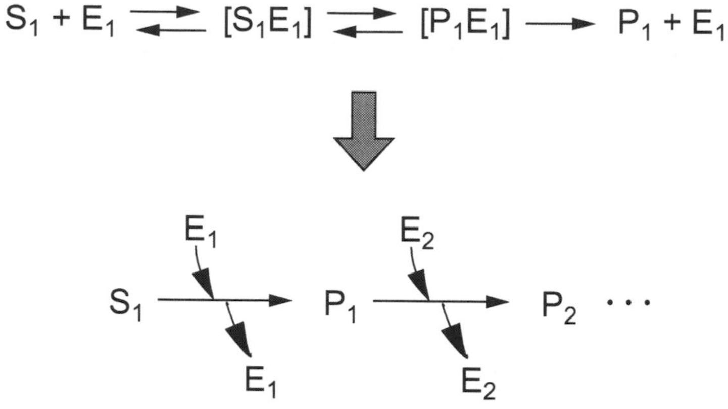

$$S_1 + E_1 \; \rightleftharpoons \; [S_1E_1] \; \rightleftharpoons \; [P_1E_1] \; \longrightarrow \; P_1 + E_1$$

FIGURE 4.1: A schematic diagram of a metabolic pathway. The three reactions at the top—enzyme E_1 and substrate S_1 forming a complex, conversion of substrate into product P_1, followed by dissociation of E_1 and P_1—are represented in a metabolic network as a single "meta-edge" depicting the conversion of substrate S_1 into product P_1, which, in turn, acts as a substrate for the next reaction in the pathway.

zyme phenylalanine hydroxylase (PAH), which catalyzes the conversion of the amino acid phenylalanine to the amino acid tyrosine. Tyrosine is involved in the production of the pigment melanin in the skin, but more importantly, when excess levels of phenylalanine build up in the body, it is converted to phenylpyruvate, which is implicated in brain damage. This condition is known as phenylketonuria (PKU). Thus, mutations in the PAH gene (or translation errors in the production of the PAH enzyme) lead to low levels of melanin as well as to PKU. These processes are described by the metabolic pathways for the conversion of phenylalanine to tyrosine and the conversion of phenylalanine to phenylpyruvate. Knowledge of the reactions in these pathways and the rate constants governing them allows one to quantitavely predict the phenotypic effects of reduced PAH expression.

There is extensive available literature on the analysis of metabolic networks [see, for example, the book by Palsson (2006)]. Particular attention has been traditionally paid to the computational technique of metabolic network analysis known as Flux Balance Analysis (FBA), which does not require detailed knowledge of pathway/network parameters. Our foray into metabolic networks here is motivated by their special role in enabling the genotype-phenotype connection—in particular, in the elucidation of genetic interactions.

Box 4.1 Chemical reaction parameters.

Consider a chemical reaction in which one molecule of chemical species A dissociates into m molecules of species B and n molecules of species C. This is generally a reversible reaction, represented as

$$A \underset{k_r}{\overset{k_f}{\rightleftharpoons}} mB + nC,$$

where k_f and k_r are *rate constants* that determine the rates of the forward and reverse reactions, respectively, and m and n are *stoichiometric coefficients*. According to the *law of mass action*, which is valid for single-step reactions and sufficiently large concentrations, the rate of the forward reaction is $k_f[A]$ and the rate of the reverse reaction is $k_r[B]^m[C]^n$, where $[A]$ and $[B]$ are the concentrations of A and B. These rates are also called *reaction fluxes*. At equilibrium, the forward and reverse reaction fluxes are equal, keeping the concentration of each chemical species constant. Thus, at equilibrium, we have

$$k_f[A] = k_r[B]^m[C]^n,$$

and the *equilibrium constant*, defined as $k_{eq} = k_f/k_r$, is

$$k_{eq} = \frac{[B]^m[C]^n}{[A]}.$$

The equilibrium constant has an important thermodynamic interpretation: it is related to the change in the Gibbs free energy ΔG when A dissociates into B and C at equilibrium, as

$$\Delta G = -RT \ln k_{eq},$$

where R is the gas constant (approximately 8.31 J/(mol-K)) and T is the temperature in Kelvin. In this book, the term "reaction parameters" generically denotes rate constants, equilibrium constants, and species concentrations.

4.1 Optimal Metabolic States and Flux Balance Analysis

Even though the topology or wiring diagram of metabolic networks remains unchanged over timescales corresponding to cellular lifetimes (in fact, the topology of important metabolic networks is strongly conserved across organisms, suggesting that it remains largely unchanged even over evolutionary timescales), the actual *state* of the network is generally time dependent because species concentrations are dynamic. As we have discussed, the behavior of the network is regulated at the genome level by the regulation of expression of the enzymes that catalyze each reaction in the network. Furthermore, the state of the network generally has clear phenotypic implications.

The genome determines the parts list of metabolic enzymes. The regulatory network, in principle, determines the expression levels/concentrations of these enzymes. The individual metabolite concentrations inside the cell are partly determined by the external concentrations of a few specific metabolites (which is usually ignored by assuming an unlimited supply of these). The numerical values of the enzyme concentrations, coupled with knowledge of the metabolic network, can be used to computationally simulate the reactions of metabolism, and thus to predict the phenotypic "states" that correspond to the metabolic network states.

More importantly for our purposes, the phenotypic effects of metabolic gene deletions (leading to nonexpression of metabolic enzymes) can be studied *in silico* by assuming that the reactions those enzymes catalyze are "turned off." Similarly, by studying the phenotypic effects of pairs of gene deletions, we are then able to predict genetic interactions among metabolic genes. This program for predicting the effects of metabolic gene deletions, straightforward as it seems, is far from trivial to implement and is also limited by the amount of available information. As previously indicated, a serious bottleneck in the program is the lack of knowledge of chemical reaction parameters that govern the rates of metabolic reactions. These parameters include rate constants that are, in turn, determined from more fundamental parameters such as enzyme-substrate binding energies as well as other factors.

While there are *in vitro* measurements of many of these parameters, it is not clear whether these values are applicable for *in vivo* calculations. Furthermore, even if reaction parameters are known, computational simulation of large metabolic networks is a daunting task involving simultaneous numerical solutions of a large system of coupled nonlinear ordinary differential equations. An even higher order of complexity is evident when one imagines trying to hunt down genetic interactions by systematically carrying out simulated deletions of every possible pair of genes and solving the entire set of associated differential equations for each deletion.

Box 4.2 Chemostats.

For experimental purposes, it is desirable to create an environment for microorganisms to grow in which their growth rate can be externally controlled. Such an environment is provided by the *chemostat*. A chemostat is essentially a chemical reactor that contains microorganisms living in a nutrient-rich medium. The volume in the reactor is kept constant by adding medium and removing culture at appropriate rates as the microorganisms grow. The important aspect of a chemostat is that the growth rate of the microorganisms can be controlled by controlling the rate at which nutrient-rich medium is added to the chemostat (and consequently, the rate at which culture is removed). The medium typically contains at least one nutrient that is essential for growth. By supplying this nutrient at limiting, rather than surplus, concentrations, the rate of growth of microorganisms can be tightly controlled and steady state (constant volume, constant nutrient concentration, constant cell density, etc.) can be achieved.

4.1.1 Flux Balance Analysis: Basic Assumptions

Fortunately, and somewhat surprisingly, it turns out that the *steady-state* behavior of a metabolic network can be efficiently determined to reasonable accuracy without knowledge of reaction parameters and without having to efficiently simulate large systems of coupled nonlinear differential equations.

The basic assumption is that the steady state of a metabolic network is *optimal* in some sense, that is, the chemical reaction fluxes are distributed in such a way as to maximize or minimize some biologically important property that can be mathematically expressed as a function of the fluxes. This property may vary depending on the environmental conditions. The computational problem then involves finding the flux distribution that achieves the desired maximum or minimum, regardless of the reaction parameters. Examples of biologically important optimizations include maximization of biomass formation (leading to maximal cell growth rate), minimization of ATP production in mitochondria (leading to the efficient use of energy), minimization of nutrient uptake in chemostats (leading to the efficient use of nutrients; see Box 4.2 for a description of the chemostat environment), and maximization of production of a specific metabolite (this has different interpretations depending on the metabolite in question).

How does one mathematically optimize a phenotypic property to in-

fer the steady-state fluxes in a metabolic network? This is the subject
of metabolic *flux balance analysis* (FBA), a topic that has been extensively studied and written about, including a detailed exposition by
Palsson (2006). Here, we focus on the main ideas that render such an
optimization problem feasible.

The rate of change of concentration of each metabolite in a metabolic
network can be represented in the form of a simple ordinary differential
equation of the form

$$\frac{dX_i}{dt} = V_{\text{syn}} - V_{\text{deg}} - V_{\text{use}} + b_i, \tag{4.1}$$

where X_i is the concentration of metabolite i; V_{syn} and V_{deg} are the
total rates of synthesis and degradation of metabolite i, respectively;
V_{use} is the rate at which metabolite i is used or consumed for metabolic
requirements; and the last term b_i is the net rate of transport of metabolite i into the cell. This last term is only relevant for metabolites that
flow across the cell boundary. Each total rate V_{syn}, V_{deg}, and V_{use} is,
in turn, a linear combination of individual reaction fluxes, so that if
we collect together the concentrations of all metabolites into a single
vector $\mathbf{X}$, the rate of change of $\mathbf{X}$ can be written as

$$\frac{d\mathbf{X}}{dt} = \mathbf{S} \cdot \mathbf{v} + \mathbf{b}, \tag{4.2}$$

where $\mathbf{v}$ is the set of individual reaction fluxes and $\mathbf{b}$ is the vector of
transport rates to and from the extracellular medium. $\mathbf{S}$ is an $m \times n$
stoichiometry matrix, where m is the number of metabolites, and n is
the number of reactions. A brief discussion of the stoichiometry matrix
description of chemical kinetics is given in Box 4.3.

Note that a system of metabolic reactions incorporates genomic information through the enzymes that catalyze each reaction. These enzymes affect the individual reaction fluxes but not the overall stoichiometry matrix (see Box 4.3).

It is usually convenient to redefine the vector $\mathbf{v}$ of reaction fluxes so
as to include the transport rates $\mathbf{b}$, and correspondingly augment the
stoichiometry matrix appropriately (Box 4.3). This yields

$$\frac{d\mathbf{X}}{dt} = \mathbf{S}' \cdot \mathbf{v}', \tag{4.5}$$

where $\mathbf{v}'$ includes *all* fluxes, including those representing transport
across the cell boundary, and $\mathbf{S}'$ is the stoichiometry matrix augmented

Box 4.3 Reaction fluxes and the stoichiometry matrix.

Consider the simple example of reversible protein dimerization, where two protein monomers bind to form a dimer. This chemical reaction can be represented as

$$2A \underset{k_r}{\overset{k_f}{\rightleftharpoons}} B,$$

where k_f and k_r are the usual rate constants. The reversible reaction formally corresponds to two reactions: the forward and reverse reactions, with fluxes

$$v_f = k_f[A]^2 \quad \text{and} \quad v_r = k_r[B],$$

respectively. Thus, the rates of production of A and B are given by

$$\frac{d[A]}{dt} = -2v_f + 2v_r, \quad \frac{d[B]}{dt} = v_f - v_r, \tag{4.3}$$

where the factor of 2 in the first equation arises from the stoichiometry of A and ensures that $[A] + 2[B]$ always remains a conserved quantity, as it should be. The two equations above can be written as a single equation in matrix form by defining a vector of species concentrations $\mathbf{X}$, a *stoichiometry matrix* $\mathbf{S}$ and a vector of fluxes $\mathbf{v}$ as follows:

$$\mathbf{X} = \begin{pmatrix} [A] \\ [B] \end{pmatrix}, \quad \mathbf{S} = \begin{pmatrix} -2 & 2 \\ 1 & -1 \end{pmatrix}, \quad \mathbf{v} = \begin{pmatrix} v_f \\ v_r \end{pmatrix}.$$

Then, Equations (4.3) can be written as the single matrix equation

$$\frac{d\mathbf{X}}{dt} = \mathbf{S} \cdot \mathbf{v}, \tag{4.4}$$

where $\cdot$ denotes ordinary matrix multiplication. In general, $\mathbf{S}$ has m rows and n columns, where m is the number of chemical species and n is the number of reactions. The elements of the stoichiometry matrix are simply the stoichiometric coefficients of the various chemical species, augmented by a $-$ sign if the corresponding reaction depletes the chemical species in question. It is noteworthy that even if the elementary reactions are not single-step reactions and therefore do not obey mass action kinetics (such as reactions catalyzed by enzymes), the mathematical expressions for the reaction fluxes will change but the stoichiometry matrix remains the same. If chemicals are externally added to or removed from the system, there will be an additional term in Equation (4.4), as in Equation (4.2). This term can be treated as an additional flux, and the stoichiometry matrix can be suitably modified so that the full system can still be represented by an equation like Equation (4.4). In the context of our example, let us suppose that A is being added at a constant rate b_1. Then we augment $\mathbf{v}$ with an additional row with the entry b_1 and augment $\mathbf{S}$ with an additional column in which the entry in the row corresponding to A is 1 and that in the row corresponding to B is 0.

by an identity matrix. At steady state, defined by the concentrations of species being constant in time $(dX_i/dt = 0)$, we obtain

$$\mathbf{S}' \cdot \mathbf{v}' = 0. \tag{4.6}$$

The above system represents m equations in n unknown fluxes. Because the number of metabolites is usually far less than the number of reactions (each metabolite is expected to participate in more than one reaction on average), the set of fluxes is underdetermined by the above equations. In order to determine the fluxes, additional conditions must therefore be imposed.

First, we recognize that the space of solutions of Equation (4.6) is a linear subspace of the full flux space; that is, if $\mathbf{v}_a$ and $\mathbf{v}_b$ are solutions of Equation (4.6), then so is any linear combination of $\mathbf{v}_a$ and $\mathbf{v}_b$. This linear subspace is called the *null space* of the augmented stoichiometry matrix $\mathbf{S}'$. The fact that any valid set of steady-state fluxes must lie in this null space corresponds to one set of constraints on the allowed fluxes.

Other constraints on individual flux values arise from the fact that each flux has a maximum value because enzymes cannot catalyze reactions at arbitrarily high speeds. Furthermore, for fluxes corresponding to the transport of material within the cell, there are additional limitations arising due to bounds on transport rates. Similarly, each flux is expected to have a minimum value based on the expectation that the chemical species in question must be synthesized at some minimal rate (in computational models, this minimum rate can be set to zero if its value cannot be estimated with reasonable accuracy from *in vitro* studies or other measurements). These considerations lead to inequality constraints on the fluxes of the type

$$\alpha \leq v_i \leq \beta. \tag{4.7}$$

The set of inequality constraints on individual fluxes (or on linear combinations of individual fluxes) leads to the allowed null space being bounded. This so-called "capped" null space is schematically represented in Figure 4.2. Every point inside the capped null space is an allowed steady-state solution, while the "caps" or boundaries arise because of the imposed lower and upper limits on the individual fluxes.

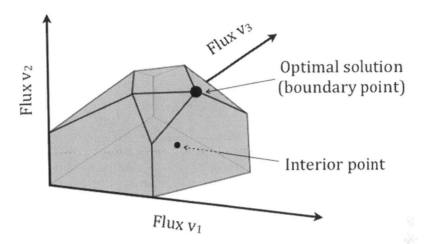

FIGURE 4.2: A schematic diagram of the capped null space of the stoichiometry matrix (the space of feasible flux distributions) of a metabolic network [see, for example, Varma and Palsson (1994)]. The "caps" are the plane surfaces that enclose the gray region of allowed values for the fluxes. These planar caps correspond to inequality constraints [see the discussion leading to Equation (4.7)]. The point labeled as the "optimal solution" corresponds to a set of fluxes that maximize some objective function. Such a point must lie on the boundary of the capped null space rather than in the interior, as argued in the text.

4.1.2 Flux Balance Analysis: The Objective Function

The aim of FBA is to find a particular steady-state solution (i.e., a point inside the capped null space) that corresponds to optimization of a biologically important property, as mentioned above. Such properties are mathematically termed *objective functions* of interest, and one of the most important objective functions is one that describes the rate of growth of the cell. At least in prokaryotes, it is reasonable to conjecture that cells have evolved to maximize their growth rate. Because one of the functions of a metabolic network is to synthesize the entire *biomass* composition of the organism, we can hope to find the steady-state distribution of fluxes in a metabolic network that maximize the production of biomass (and hence, its growth rate). This problem is best addressed by introducing a new flux v_{growth} describing the conversion of certain important metabolites (called *biosynthetic precursors*) into biomass. This reaction can be represented as

$$\sum_m c_m X_m \xrightarrow{v_{\text{growth}}} \text{biomass}, \qquad (4.8)$$

where the stoichiometric coefficients c_m specify relative proportions of various metabolites in biomass composition, and differ across organisms and even across growth environments. Values of these coefficients are measured and documented for many organisms, for example the bacterium *Escherichia coli*, whose metabolic properties have been extensively studied (Neidhardt et al., 1996). For other organisms, such as *Helicobacter pylori*, the coefficients of biomass compositions are not available because these organisms cannot be grown under defined and controlled nutritional conditions.

4.1.3   Optimization Problem in Flux Balance Analysis

In summary, the optimization problem of interest consists of maximizing one of the fluxes (such as the growth flux above) subject to the constraints of a steady state and the inequality constraints corresponding to enzymatic capacity limitations. In other words, we wish to find a single point within the capped null space that corresponds to the maximum value of the growth flux. Mathematically, this type of problem is called a *linear programming* (LP) problem: it is characterized by an objective function that is a linear function of the variables (in this case, the variables are the fluxes), and by constraints that are either linear equality constraints [as in Equation (4.6)], or linear inequality constraints [the capacity constraints of Equation (4.7)].

Consider a general linear programming problem where the objective function is a linear combination of the fluxes. Suppose $(v_1, v_2, v_3, \ldots)$ represents a point in the interior of the capped null space, as shown in Figure 4.2. Let us assume that the value of the objective function at this interior point is positive. Because the objective function is linear in the fluxes, multiplication of *every* flux by a common number that is larger than 1 will result in the objective function being multiplied by the same factor. Thus the objective function will increase in value. Therefore, it is always possible to increase the value of the objective function at an interior point of the capped null space by moving to a neighboring point where all the fluxes are somewhat larger in magnitude. It follows that the maximum value of the objective function cannot be attained at an interior point of the capped null space, but rather at a boundary point[1], as illustrated in Figure 4.2.

For example, *Escherichia coli* needs various metabolites in the right proportion to grow biomass. Two of the most important metabolites are ATP (adenosine triphosphate) and NADH (reduced form of nicotinamide adenine dinucleotide). In order to increase *E. coli* biomass by 1 gram, it turns out that 41.257 mmol ATP (as the molecular weight of ATP is 507.18 g/mol, this corresponds to 20.925 g ATP) must be added to the growth medium, while 3.547 mmol NADH (corresponding to 2.353 g NADH) must be removed from the growth medium. Various other metabolites must also be added in the right proportions to effect the 1-g biomass increase, but for now let us assume a simplistic model where just the two metabolites ATP and NADH need to be altered. In this simplistic model, the growth objective function is

$$Z = 41.257\, v_{\text{ATP}} - 3.547\, v_{\text{NADH}}. \tag{4.9}$$

It is clear that the above function attains its maximum value when the flux v_{ATP} takes its maximum allowed value subject to the constraints that we have discussed and when the flux v_{NADH} takes its minimum allowed value subject to the constraints. These maximum and minimum flux values are obviously attained at a boundary point of the capped null space.

The simple arguments above show that solutions to linear programming problems always lie at the boundaries or edges of the feasible space. The aim of flux balance analysis is to find the boundary points corresponding to steady-state fluxes that maximize biomass production. Remarkably, these fluxes reproduce many of the essential features of metabolism in prokaryotes, indicating that, to reasonable approximation, prokaryotes have indeed evolved to maximize cell growth.

Given a linear objective function and a set of inequality constraints on the fluxes, a simple way to computationally implement linear programming for a large metabolic network, and thus to optimize the objective function, is via the use of the COBRA software (Becker et al., 2007; Orth et al., 2010).

4.1.4   FBA Enables the Genotype–Phenotype Connection

The FBA paradigm is a way of incorporating genomic information (e.g., expression levels of metabolic enzymes) into the prediction of pheno-

---

[1]If the objective function at an interior point is negative, then it can be increased by multiplying all fluxes by a factor less than 1, and a similar argument applies.

typical behavior, for example, the rate of biomass formation or cell growth. It follows that this paradigm may be used to predict phenotypical changes that occur as a result of perturbations to the genotype. These perturbations can be carried out on the computer in various ways. Here are some scenarios.

- The effect of knockout of a metabolic gene may be simulated by setting the flux of the reaction that the corresponding enzyme catalyzes to zero. This constraint can be incorporated into the linear programming problem and the resulting change in the rate of biomass formation computed. More accurate predictions of phenotypical change are possible if the MOMA (minimization of metabolic adjustment) procedure (discussed in the next section) is used as an alternative to FBA.

- By altering the lower and upper bounds on the speed of certain reactions, one can study the genotype-phenotype connection under various conditions.

- *Equality* constraints can be handled in the linear programming framework as special cases of inequality contraints corresponding to setting the lower and upper bounds equal to each other. This is useful because one can study the behavior of metabolic networks under conditions where one or more fluxes are held fixed. This can even happen in a dynamic fashion. For example, if the flux through a certain reaction is repressed for a certain period of time because of transcriptional regulation of the associated enzyme during that time interval, this can be represented by setting the corresponding flux to a small value that is constant over the time interval (Covert et al., 2001). The optimization problem then must be re-solved every time the constraint changes.

- The sensitivity of phenotypic changes to individual reaction fluxes can be found by changing each flux by a small amount in a controlled fashion and measuring the corresponding change in phenotype.

4.2 Minimization of Metabolic Adjustment and Gene Deletions

As we have outlined, one of the central tenets of the FBA paradigm is that the natural flux distribution corresponds to the optimal solution of some objective function that is linear in the fluxes. Typically, this function is the growth rate of the organism, as it is reasonable to hypothesize—at least for simple unicellular organisms—that the organism has evolved toward maximal growth performance. While this assumption has been shown to (quite remarkably!) hold for wild-type prokaryotes, the situation for organisms that suffer gene knockouts is far from clear. It is not necessary that these artificially engineered mutants must also optimize their growth rate as they do not correspond to naturally evolving populations. Thus, results obtained from the application of the FBA principle to metabolic network models in which a gene knockout has been incorporated must be considered suspect.

For computing flux distributions for organisms that undergo gene knockouts, Segré et al. (2002) developed the principle of the minimization of metabolic adjustment (MOMA) as an alternative hypothesis to FBA. Recall that a set of metabolic reaction fluxes, or flux distribution, can be represented as a point in the capped null space. In MOMA, one assumes that the flux distribution that is realized following one or more gene knockouts is closest, or has the shortest distance, to the point corresponding to the wild-type flux distribution in the capped null space, rather than being a point that is optimal for a phenotypic property such as growth rate (as FBA assumes). Thus the central assumption of MOMA is that a gene deletion shifts the flux distribution from the optimal one to the closest realizable flux distribution. This closest distribution is generally different from the original optimal one because the optimal distribution may no longer be realizable under the additional constraint corresponding to deletion of the gene in question.

Because distance between two flux distributions is related to a quadratic function of the fluxes, mathematically the problem of finding the new flux distribution after gene deletion is a quadratic programming problem rather than a linear programming one. The resulting "sub-optimal" distribution of fluxes is generally quite different from the one obtained by application of FBA and, in particular, has different phenotypical implications. Segré et al. found that MOMA reproduces the true state of the "knocked-out" organism more accurately than FBA (Figure 4.3.)

Having studied how the effects of single-gene knockouts can be computationally simulated, we now turn to the question of the effect of double-gene knockouts, that is, genetic interactions. As we have seen earlier, genetic interactions directly relate physical interactions with their phenotypic consequences. For example, as we shall find out in greater detail in Chapter 9, the identification of synthetic lethal genetic interactions can be directly translated to the discovery of novel therapeutic targets or treatment regimens. The idea of using genetic interactions for therapeutic purposes has turned out to be quite promising. In certain types of cancers, two simultaneous mutations in the major DNA repair pathways make the cancer cells exquisitely dependent on a normally minor repair pathway (Figure 4.4). These cancer cells then could be precisely targeted by a combination treatment that simultaneously causes DNA damage by general chemotherapy and a targeted drug that abrogates the minor DNA repair pathway. Because non-cancer cells do not depend on the minor repair pathway to maintain the integrity of their genomes, only the cancer cells are highly susceptible. Note that the particular relationship among the three pathways is formally equivalent to that shown in the asynthetic genetic interaction panel of Figure 2.2. The allele states of the various genes in the three pathways can be ascertained by DNA sequencing of the cancer cells of the particular patient. Targeted therapy in combination with the DNA damaging agent is used against the minor pathway only if mutations in the two major pathways exist in the cancer cells but not in the healthy cells. This approach to cancer chemotherapy therefore uses patient-specific gene network based knowledge, and is an example of personalized therapy. It is anticipated that such knowledge will increasingly become important in personalized medicine and targeted therapy in human diseases.

4.3 Predicting Genetic Interactions among Metabolic Genes

As we have seen, FBA and MOMA are two methods, built upon the same conceptual foundation, that attempt to computationally simulate the effects of genetic perturbations on phenotype. Although the MOMA method more accurately captures the effect of single gene deletions, in

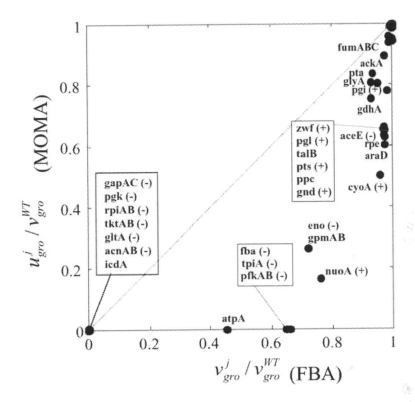

FIGURE 4.3: Comparison of growth yield as calculated by MOMA (y-axis) with that calculated by FBA (x-axis) for *Escherichia coli* mutants in which individual genes are deleted. Each point is labeled by the deleted gene, and (+) and (−) correspond to genes that are nonessential and essential, respectively. Essential gene knockouts should correctly lead to zero growth. In both axes, growth yields are normalized by their wild-type values. Thus, u_{gro}^j and ν_{gro}^j are the growth yields of the mutant as predicted by MOMA and FBA, respectively, while ν_{gro}^{WT} is the wild-type growth yield as predicted by FBA. As expected from the sub-optimality of the MOMA procedure, all points lie below the diagonal because the MOMA-predicted growth rate is less than the FBA-predicted one. Points on the lower left-hand corner denote lethal deletions identified correctly by both FBA and MOMA. Points on the top-right corner denote nonessential deletions that are also correctly identified by both procedures. In other cases, predictions of the two procedures disagree, with MOMA being more accurate on average: three essential genes are correctly identified by MOMA but not by FBA, while no nonessential gene was incorrectly identified by either FBA or MOMA (although some nonessential genes, like *nuoA*, were predicted to have low growth rates by MOMA but not by FBA). Reproduced with permission from Segré et al. (2002). [©2002. National Academy of Sciences, U.S.A.]

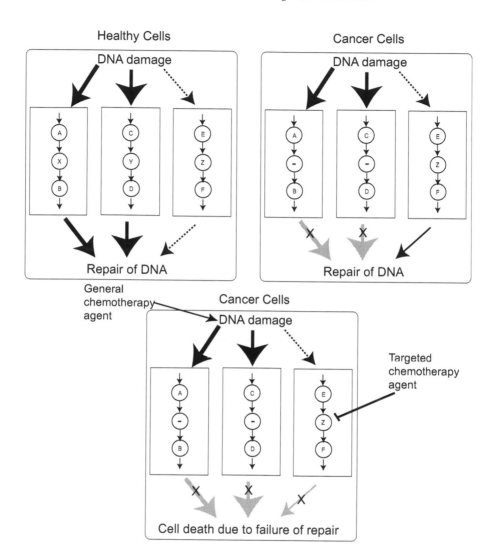

FIGURE 4.4: A rationale for personalized and targeted chemotherapy based on genetic interaction data. Healthy cells repair their damaged DNA through one or both of two major pathways: A-X-B and C-Y-D. A minor pathway, E-Z-F, also exists. In some cancer cells, DNA sequencing of the repair genes has shown that the repair genes X and Y might both be mutated, such that the normally minor pathway, E-Z-F, becomes the only available pathway of DNA repair in these cancer cells. This minor pathway can then be targeted for chemotherapeutic blockade. However, the healthy cells of the patient are not hypersensitive, because the major pathways of DNA repair, A-X-B and C-Y-D, are still functional in these cells.

this section we return to traditional FBA as a tool for analyzing the effects of multiple gene deletions because it simpler in implementation. There is no conceptual barrier to replacing the FBA-based formulation outlined below by a MOMA-based formulation.

In FBA, the phenotypic effect of a knockout can be monitored by defining a "fitness" function that measures the rate of biomass production in the mutant organism relative to that in the wild-type organism. That is, if ΔX denotes the "knockout of gene/enzyme X" and V_{growth} is the optimized rate of biomass production, then the fitness W_X of a mutant organism in which gene X has been deleted is given by

$$W_X = \frac{V_{\text{growth}}^{\Delta X}}{V_{\text{growth}}^{\text{wild-type}}}. \tag{4.10}$$

Note that in this framework the fitness of the wild-type organism is 1, by definition.

It is important to observe that the effect of a gene knockout or deletion cannot simply be modeled by setting the appropriate flux to zero *without* changing any of the other fluxes. On the contrary, setting one of the fluxes to zero is an additional constraint on the linear programming problem, which must now be solved anew, generally resulting in a completely different set of fluxes. It is true, however, that the space of fluxes that must now be searched to find the optimal solution is a subspace of the original search space, and therefore the optimal rate of biomass production as found in the presence of the additional constraint can be no higher than the optimal rate in the absence of that constraint. In other words, $V_{\text{growth}}^{\Delta X} \leq V_{\text{growth}}^{\text{wild-type}}$, and therefore $W_X \leq 1$. If we similarly define W_{XY} as the fitness (ratio of optimal growth rates) of the organism in which *both* genes X and Y have been deleted, then a similar line of reasoning implies that $W_{XY} \leq W_X$ and $W_{XY} \leq W_Y$. This argument further establishes that, within the FBA framework, the fitness of triple mutants can be no higher than that of their parent double mutants, and so on. Thus, all mutations in the FBA framework are either *neutral* (i.e., do not change fitness) or *deleterious* (i.e., decrease fitness). This phenomenon is of course an artifact of the FBA approximation: namely, the assumption that the organism (whether wild-type or mutant) lives in a state of optimality with respect to its rate of biomass production. This assumption cannot be expected to be true in nature; living organisms do not generally exist in a state of perfectly optimal growth, thus leaving room for rare, beneficial mutations that increase fitness. Further, additional mutations that "hitchhike" on deleterious mutations can sometimes prove beneficial relative to the wild-type. None of these effects are accounted for in the approximate FBA framework, although they are, to some extent, accounted for in the MOMA framework. Even so, the simplicity of the FBA framework and the relative rarity of beneficial mutations encourage us to consider seriously its predictions.

4.3.1 Computational Simulation of Mutation Effects

Segré et al. (2005) studied the computationally simulated effects of single and double gene deletions of 890 metabolic genes in *Saccharomyces cerevisiae* using the framework of FBA and the fitness function defined above. It is customary in the population genetics literature to use the term "epistasis" to mean any genetic interaction. This usage is in contrast to the terminology used elsewhere in this book (and in the molecular biology literature), where epistatic interactions constitute a specific type of genetic interaction. For clarity and consistency, whenever we wish to speak of a generic genetic interaction, we will therefore use the term "influence" instead of "epistasis." Thus, if the fitness of a mutant organism in which two genes are deleted is not the product of the fitnesses of the two single mutant organisms, we will say that the two genes genetically interact or *influence* each other. The amount of influence can be quantified by (Collins et al., 2006)

$$\epsilon = W_{XY} - W_X W_Y. \tag{4.11}$$

Non-zero values of ϵ simply imply a genetic interaction between genes X and Y, while $\epsilon = 0$ corresponds to lack of genetic interaction (a "noninteractive" case according to the terminology in Table 2.3 and Figure 2.2), or *multiplicative influence*. This terminology stems from the fact that the independent fitness effects of the two single-mutants are multiplied in the above measure; additive measures of influence are also possible, as discussed in Chapter 2, and so are statistical measures where the effect of a specific double gene deletion is compared to the average effect of an ensemble of double gene deletions. For detailed expositions of quantitative measures of genetic influence, the reader is directed to the work of Cordell (2002) as well as further work by Snitkin and Segré (2011).

The case $\epsilon > 0$ corresponds to a genetic interaction of the *buffering* type, an extreme example being the case where W_{XY} takes its largest

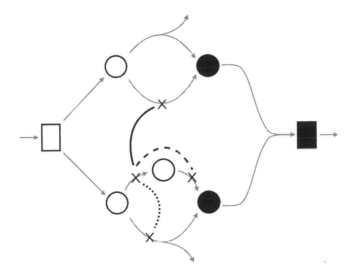

FIGURE 4.5: A schematic metabolic network showing buffering (dashed black curve), aggravating (dotted black curve), and multiplicative (solid black curve) genetic interactions. The network represents conversion of a common nutrient (empty rectangle) to a series of intermediate metabolites (empty circles) to amino acids and nucleotides (filled circles) to biomass (filled rectangle). Deletions of metabolic genes are marked by crosses ("x") placed on metabolic network edges (thin gray arrows) that signify reactions mediated by enzymes corresponding to the metabolic genes. Each genetic interaction is uncovered by the phenotype corresponding to the deletion of both members of a pair of metabolic genes. Note that the arrows at the very top and bottom of the figure correspond to metabolic conversions that will *not* lead to production of biomass. Adapted from Segré et al. (2005).

allowed value, namely $\min(W_X, W_Y)$ (the argument at the beginning of this section shows that no larger value is possible). In this extreme situation, one mutation completely buffers the effect of the other, less deleterious mutation.

The case $\epsilon < 0$ corresponds to a genetic interaction of the *aggravating* type, with an extreme example being that of the synthetic lethal interactions, where $W_{XY} = 0$, $W_X = W_Y = 1$, and $\epsilon = -1$. Figure 4.5 schematically displays the kinds of metabolic enzyme knockouts that are likely to result in multiplicative, buffering, and aggravating interactions. Note that aggravating and buffering interactions lie outside the scope of the interactions covered in Table 2.3.

Because the FBA model does not admit beneficial mutations, it turns out that ϵ has both upper and lower bounds, as discussed above. Its minimum possible value is -1, corresponding to synthetic lethal interactions, and its maximum possible value is obtained for perfectly buffering interactions. It is therefore convenient to define a rescaled influence measure as follows:

$$\tilde{\epsilon} = \frac{W_{XY} - W_X W_Y}{|\tilde{W}_{XY} - W_X W_Y|}, \tag{4.12}$$

where $\tilde{W}_{XY} = \min(W_X, W_Y)$ if $W_{XY} > W_X W_Y$, and $\tilde{W}_{XY} = 0$ if $W_{XY} \le W_X W_Y$. With this rescaling, $\tilde{\epsilon} = -1$ for synthetic lethal interactions and $\tilde{\epsilon} = 1$ for perfectly buffering interactions. After excluding essential metabolic genes and gene pairs for which neither single nor double knockouts have phenotypic effects, Segré et al. found that the rescaled influence measure exhibits a sharply tri-modal disribution, with about 600 gene pairs displaying multiplicative influence (equivalent to no influence, $\tilde{\epsilon} = 0$), about 200 gene pairs predicted to have synthetic lethal interactions, and about 100 gene pairs predicted to be nearly perfectly buffering. Remarkably, very few gene pairs are predicted to have influence values that are significantly different from these three extremes, a phenomenon that is consistent with limited experimental measurements of the fitness of RNA viruses. Thus, the influence effects of all gene pairs in baker's yeast were classified as buffering, aggravating, or multiplicative.

Segré et al. also found that when genes were clustered into functional categories based on the metabolic process they participated in, the set of genetic interactions between all genes in one functional class and all genes in another functional class were predominantly "monochromatic," that is, they were either all aggravating or all buffering (this was found to be true even for all genetic interactions within the same functional class), with few exceptions (Figure 4.6). This second prediction is consistent with the picture of a gene being defined mainly by its functional role. Indeed, it was found that the exceptional non-monochromaticity is more prevalent among genes that belong to more than one functional category.

Remarkably, the feature of monochromaticity of genetic interactions survives even when the computational study of the effects of double gene deletions is analyzed with respect to phenotypes other than biomass production or cell growth. Snitkin and Segré (2011) considered the effects of these deletions on 80 different metabolic phenotypes (here MOMA, rather than optimization of biomass growth, was used

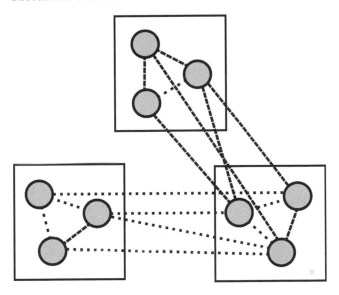

FIGURE 4.6: Monochromaticity of genetic interactions predicted by the FBA model. As in the previous figure, dotted and dashed lines represent aggravating and buffering genetic interactions, respectively. Nodes represent genes; genes that belong to the same functional class are displayed within the same box. Monochromaticity refers to the situation in which genetic interactions between genes in different functional classes (different boxes) are either all aggravating or all buffering (or nearly so), although interactions within a functional class can be either aggravating or buffering.

to calculate the modified fluxes), and found that genetic interactions between pairs of biological processes, for all phenotypes taken together, were mostly all aggravating or all buffering, although gene pairs often interact in an aggravating manner relative to some phenotypes and buffering with respect to others (Figure 4.7).

4.4 General Methods for Predicting Genetic Interactions

As we have seen, the FBA and MOMA methods are valuable in their ability to computationally simulate the phenotypic effects of gene knockouts and hence predict genetic interactions. However, they are limited by the fact that only metabolic gene knockouts can be simulated for examining their potential phenotypic effects. Because the

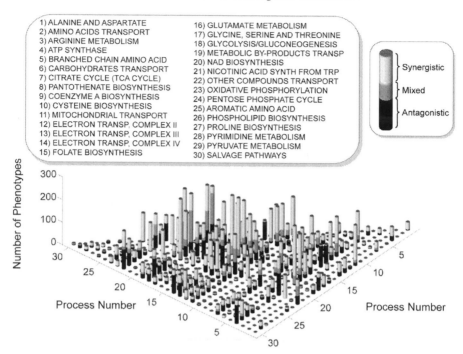

1) ALANINE AND ASPARTATE
2) AMINO ACIDS TRANSPORT
3) ARGININE METABOLISM
4) ATP SYNTHASE
5) BRANCHED CHAIN AMINO ACID
6) CARBOHYDRATES TRANSPORT
7) CITRATE CYCLE (TCA CYCLE)
8) PANTOTHENATE BIOSYNTHESIS
9) COENZYME A BIOSYNTHESIS
10) CYSTEINE BIOSYNTHESIS
11) MITOCHONDRIAL TRANSPORT
12) ELECTRON TRANSP, COMPLEX II
13) ELECTRON TRANSP, COMPLEX III
14) ELECTRON TRANSP, COMPLEX IV
15) FOLATE BIOSYNTHESIS

16) GLUTAMATE METABOLISM
17) GLYCINE, SERINE AND THREONINE
18) GLYCOLYSIS/GLUCONEOGENESIS
19) METABOLIC BY-PRODUCTS TRANSP
20) NAD BIOSYNTHESIS
21) NICOTINIC ACID SYNTH FROM TRP
22) OTHER COMPOUNDS TRANSPORT
23) OXIDATIVE PHOSPHORYLATION
24) PENTOSE PHOSPHATE CYCLE
25) AROMATIC AMINO ACID
26) PHOSPHOLIPID BIOSYNTHESIS
27) PROLINE BIOSYNTHESIS
28) PYRIMIDINE METABOLISM
29) PYRUVATE METABOLISM
30) SALVAGE PATHWAYS

Synergistic

Mixed

Antagonistic

FIGURE 4.7: Persistence of monochromaticity of genetic interactions across multiple phenotypes, as predicted by the MOMA model. Synergistic interactions are generalizations of the aggravating interactions for multiple phenotypes, while antagonistic interactions broadly refer to the buffering interactions. For each pair of biological processes, the fraction of phenotypes relative to which interactions can be either synergistic or antagonistic (i.e., mixed) is relatively low. Adapted with permission from Snitkin and Segré (2011).

majority of genes in an organism do not code for metabolic enzymes, these approaches are not feasible for discovering genetic interactions among the majority of genes in an organism.

To computationally discover gene interactions in a more general context, an indirect approach is necessary. Specifically, one looks for gene properties, called *features*, that *correlate* with the presence of genetic interaction, and uses the presence or absence of these properties to predict whether a genetic interaction is likely between the two genes in question. This prediction task is all the more challenging because of the sparsity of genetic interaction networks. For example, in yeast it is estimated that, roughly speaking, only one in fifty gene pairs will partake in a synthetic sick or lethal (SSL) genetic interaction.[2] Accurate prediction of such sparse networks requires extremely low false positive rates (Box 4.4). Controlling false positive rates in a computational method designed to predict these networks then requires that each predicted interaction have multiple sources of evidence, or multiple features. Some of the features found to be predictive of SSL interactions are shown in Table 4.1.

Because synthetic lethality is expected to be more prevalent among pairs of genes that have functional redundancy, many of the properties in Table 4.1 measure functional redundancy in different ways, and are related to the properties used to predict protein-protein interactions (which are also more prevalent among functionally similar proteins). Direct evidence for functional similarity [similar functional classifications according to the Gene Ontology (GO) or the Munich Information center for Protein Sequences (MIPS), similar structural role or enzymatic activity, similar phenotype of knockout mutant] is expected within synthetic lethal pairs.

Indirect evidence for functional similarity may also be important for predicting genetic interaction. For example, if the genes "co-occur" in the same set of organisms, that indicates the operation of an evolutionary constraint that requires both genes to be present if any one is, and hence indicates a putative functional relationship. If the protein products have direct physical interaction, participate in the same protein complex, are homologous, and/or localized, that too is indicative of similar function and therefore higher propensity for synthetic lethal interaction. However, some properties that do not obviously correlate with functional similarity, such as mutual clustering coefficient, protein interaction network position, and 2hop properties, need further clarification.

The mutual clustering coefficient is a mathematical measure of the extent to which two genes share a common set of interaction partners. The larger this measure, the more likely it is for the two genes to interact with each other. This coefficient can be defined in various ways [see Goldberg and Roth (2003) for a summary of definitions of this measure]. The "positions" of the products of the two genes in a protein

---

[2]Synthetic sick interactions are milder versions of synthetic lethal interactions: while a synthetic lethal interaction between two genes will result in cell death when both genes are knocked out, a synthetic sick interaction will only result in a lower rate of cell growth under the same conditions.

Box 4.4 False positive rate and accuracy in the prediction of sparse networks.

Suppose a computational method is designed to address the *binary classification* problem, which is the problem of predicting the correct class out of two possible classes that each datum may belong to. It is customary to call one class the *positive* (+) class and the other class the *negative* (−) class. An example that is of direct interest for our purposes is the problem of predicting whether a pair of genes interacts (+) or does not interact (−), which is a binary classification problem. One measure of prediction error in binary classification problems is the false positive rate (FPR), which is the proportion of true negatives in the data that are incorrectly predicted to be positive. Similarly, one can also define a false negative rate (FNR), which is the proportion of true positives that are incorrectly predicted to be negative.

Now let us suppose that we are trying to predict interactions between genes, and that these interactions in reality form a very sparse network, that is, the number of true positives is much smaller than the number of true negatives. For concreteness, we will assume a sparseness of one positive for every fifty negatives, which is comparable to the true sparseness of the SSL network in yeast. Furthermore, let us imagine that our prediction method is quite accurate and achieves an FNR of 0% and an FPR of 1%, that is, all true positives are correctly identified and only 1 in every 100 true negatives are incorrectly predicted to be positive. However, because of the sparseness, there are only 2 true positives for every 100 negatives. Our "accurate" prediction method will therefore predict 3 positives for every 100 negatives: it will correctly predict the 2 true positives and will additionally, and incorrectly, predict 1 false positive. Thus, this prediction method will predict 50% more interactions than actually exist, making it a very poor prediction method in spite of having zero FNR and low FPR. The catch here is to recognize that good accuracy in network prediction is only possible if the FPR is far lower than the sparseness level.

The above considerations can be formalized into the definitions of different measures of accuracy, namely the *positive and negative predictive values* (PPV and NPV, respectively). The PPV is the proportion of predicted positives that are true positives, and the NPV is the proportion of predicted negatives that are true negatives. In the above example, while the NPV is 100%, the PPV is 66.7% because only two of every three predicted positives are true positives. In general, the relationship between PPV, NPV, FNR, FPR, and the sparseness can be expressed using Bayes' rule as

$$PPV = \frac{s\,(1 - FNR)}{s\,(1 - FNR) + FPR},$$

$$NPV = \frac{1 - FPR}{1 - FPR + s\,FNR},$$

where the sparseness s is the ratio of true positives to true negatives.

TABLE 4.1: Features that Correlate with the Presence of SSL Interaction between Two Genes

Presence of a common transcription factor that binds to an
   upstream region (common upstream regulator)
Occurrence of the two genes or their homologs in the same
   set of organisms (gene co-occurrence)
Within-genome chromosomal distance (for prokaryotes)
Fusion of orthologous genes in another genome
   (gene fusion)
Chromosomal proximity of orthologous genes in other species
   (conserved gene neighborhood)
Protein-protein interaction between the gene products
   (direct and network position)
Same predicted physical complex
mRNA co-expression
Similarity of biological functions of the protein products
   (including similarity of biochemical activity and structural role)
Similarity of phenotype of knockout mutants
Protein sequence homology
Mutual clustering coefficient
Same subcellular localization
2hop properties (see Section 4.4.1)

Source: Wong et al. (2004).

interaction network can also be quantified in various ways, including
the shortest distance between the two proteins along the network, their
degrees, their betweenness centralities, and so on [see Paladugu et al.
(2008) for a summary of different measures of network position]. We
now focus on describing the 2hop properties, as these appear to correlate very well with the presence of an SSL interaction.

4.4.1 "2hop" Properties and Genetic Interactions

A "2hop" property of a pair of genes A and B is one that depends
on the existence of a third gene, C, that is related, possibly in several
different ways, to A and B. For example, it may be known that A and
C have an SSL interaction, and so do B and C. Then we say that the
"2hop S-S" property for the A-B pair has the value 1. If no such gene
C existed, we would say that the 2hop S-S property has the value 0
for the A-B pair. Figure 4.8 shows an example where the 2hop H-P
property for the A-B pair equals 1.

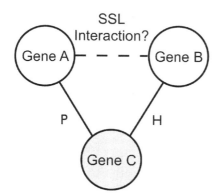

2hop H-P = 1

SSL
Interaction?

Gene A - - - Gene B

P   H

Gene C

H = Protein sequence homology
P = Protein-protein interaction
S = SSL interaction
R = Common upstream regulator
X = Correlated mRNA expression

FIGURE 4.8: Illustration of 2hop properties as indicators of the presence of SSL interactions between genes A and B. In the example shown, C has a protein-protein interaction with A and protein sequence homology with B, thus yielding 2hop H-P = 1 for the A-B pair. Other 2hop properties may be constructed using any two features from the list shown on the right. Adapted with permission from Wong et al. (2004).

The question that arises is: Given that a 2hop property equals 1 for the A-B pair, does that enhance the likelihood that an SSL interaction exists between genes A and B?

Wong et al. (2004), who introduced 2hop properties for the problem of predicting SSL interactions, answered this question in the affirmative for the 2hop properties generated by the five properties shown in Figure 4.8. Specifically, they found that 2hop P-S, 2hop S-S, and 2hop S-X are the 2hop properties that are most predictive of an SSL interaction between A and B. The fact that 2hop S-S is a strong indicator of SSL interaction is plausible given the preponderance of triangles in the known SSL network of yeast: the "functional redundancy" interpretation of SSL interactions suggests that if A and C have an SSL interaction, and so do B and C, then there is a relatively high likelihood of SSL interaction between A and B.

Similar arguments reinforce the value of other 2hop properties for SSL prediction: every property used to construct a 2hop property is an indicator of some aspect of functional redundancy.

4.4.2 Predicting Genetic Interactions from Protein Interaction Networks

In the protein interaction networks of yeast, and some other organisms, it is well known that the *degree*, or number of interacting partners, of a protein, is well correlated with the essentiality of the corresponding gene for the organism's survival. By analogy, because existence of an SSL interaction between a pair of genes is indicative of the joint essentiality of the gene *pair*, it is plausible that network properties of both of the corresponding proteins, for example their degrees, correlates with existence of SSL interaction.

Paladugu et al. (2008) examined this question in yeast by studying the predictive value of a number of protein interaction network properties of the two proteins in question for SSL interaction between the two corresponding genes. It turns out, at least in yeast, that protein interaction network properties, especially in combination with 2hop S-S and 2hop S-P, are strongly and robustly predictive of SSL interaction (Figure 4.9). In fact, Paladugu et al. also found that novel predicted SSL pairs in yeast were highly likely to possess functional similarity and common upstream regulators, even though functional and regulatory information was not used in their method for SSL prediction.

4.5 Approaches for Integrated Prediction of Networks

Network data are most meaningful when they are combined with other types of high-throughput data, such as sequence data, gene expression, protein expression, protein localization, evolutionary data, phylogenetic analysis, and functional data. We have already seen data integration at work at various points so far. Examples include the use of gene expression data to infer regulatory interactions; the use of structural features, gene neighborhood analysis, gene fusion, and evolutionary information to infer protein-protein interactions (in this context, we also discussed in Chapter 3 the integration of data using Bayesian networks); the use of metabolic networks to infer the effects of gene deletions and genetic interactions; and integration of data from various high-throughput sources to infer synthetic genetic interactions.

An example of massive data integration, one that perhaps points the way toward much of the future work in computational network biology, is provided by the development of an integrated framework to simulta-

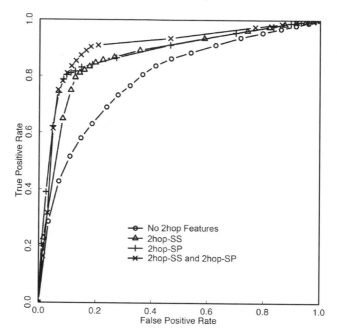

FIGURE 4.9: Accuracy of predicting SSL interaction among two genes from protein interaction network properties of the two corresponding proteins, and from 2hop properties. The y-axis is the proportion of SSL interactions that are correctly recovered at a certain prediction threshold (true positive rate), and the x-axis is the proportion of non-SSL interactions that are incorrectly predicted to be SSL at that prediction threshold (false positive rate). These curves are called Receiver Operating Characteristic (ROC) curves because of their origins in signal processing, a branch of electrical engineering. The area under an ROC curve is a measure of the overall accuracy of the prediction method. Here, the area under the curve that includes both 2hop features is about 0.9, indicating a roughly 90% average accuracy for the prediction of SSL interactions from protein interaction network properties and 2hop properties. Reproduced with permission from Paladugu et al. (2008).

neously predict different types of networks from a compendium of data sources, including information gathered from existing networks (Park et al., 2010). Such a framework integrates the predicted data in addition to the input data and is potentially very powerful because it allows the unbiased simultaneous analysis of different types of interaction, each type having been predicted using the same computational method and sharing similar error rates.

Here is a brief sketch of the prediction framework of Park et al.

that skims over many of the mathematical and computational details. First, an ontology of thirty different biomolecular interaction types was created by combining information from a number of post-genomic databases. In this ontology, certain interaction types depend on others. The thirty interaction types and the dependency relationships among them are shown in Figure 4.10. Each interaction type was independently predicted for each pair of genes in yeast by integrating data from about 3,500 experimental conditions involving expression, co-localization, regulation, and so on, with the caveat that data directly pertaining to a given network type were withheld from prediction of that type. For each gene pair, the propensities for each interaction type obtained from the prediction task are represented as scores Y_i, $i = 1, \ldots, 30$. Suppose the true interaction label for the gene pair in question is represented by the binary score X_i, where $X_i = 0$ if the two genes do not possess interaction i, and $X_i = 1$ if they do. From training data, the probability distribution $P(Y_i \mid X_i)$ can be estimated. This distribution is simply a measure of the error in predicting interaction i.

The next step is to integrate prediction of the different interaction types so that they are consistent with each other and consistent with existing data. This step uses the dependency information shown in Figure 4.10 and existing network data. Thus, if interaction i in Figure 4.10 has child interactions j, k, l, the conditional probability $P(X_i \mid X_j, X_k, X_l)$ is found by querying existing network data. For example, $P(X_i = 1 \mid X_j = 0, X_k = 1, X_l = 1)$ is estimated as the proportion of gene pairs that possess interactions k and l but do not possess interaction j, that possess interaction i. All this information can now be integrated to find the probability that the gene pair in question possesses interaction i. Denoting the set of child interactions of interaction i in the ontology tree by C_i, this probability can be expressed as

$$P(X_i \mid Y_i, C_i) \propto P(Y_i \mid X_i, C_i)\, P(X_i \mid C_i),$$
$$= P(Y_i \mid X_i)\, P(X_i \mid C_i), \qquad (4.13)$$

where the second equality follows because the predicted propensity Y_i is not dependent on the existence of child interactions. Both $P(Y_i \mid X_i)$ and $P(X_i \mid C_i)$ can be estimated from training data as outlined above, and these estimates are combined to yield $P(X_i \mid Y_i, C_i)$, which represents a modified, probabilistic score for the existence of interaction i.

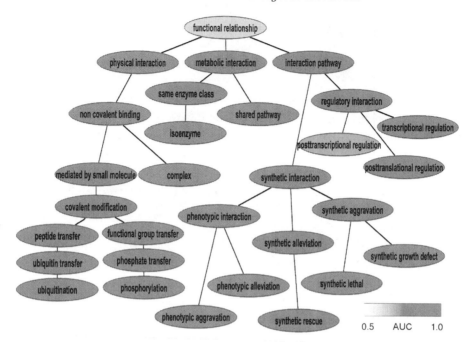

FIGURE 4.10: A tree representing the ontology of interactions. Each node in the ontology is shaded according to the independent accuracy in predicting that interaction, as quantified by the area under an ROC curve (denoted by AUC). Reproduced with permission from Park et al. (2010).

It turns out that this modified score is more accurate than the score Y_i obtained without consideration of the dependencies between various interaction types, and this is true for every interaction type studied (Park et al., 2010). The integrated database consisting of thirty interactomes can also be mined to reveal new and interesting predicted biological mechanisms.

As a case in point, Park et al. (2010) discovered a higher-order relationship between different components of the protein-sorting apparatus. Proteins are synthesized within the cytoplasm and are often enveloped within special membranous compartments (the protein cargo) from which different classes of proteins are sorted and transported to different intracellular compartments, such as the Golgi apparatus, vacuoles, plasma membrane, endoplasmic reticulum, nucleus, and so on. The heterogeneous information integration strategy led to the prediction of several genetic interactions among components that were known to occur within separate compartments. Thus the predictions provided hypotheses that, if tested, could provide novel information on how certain components could also function across compartments rather than only within compartments. Several of these predictions were validated; thus, a more integrated understanding of protein-sorting mechanisms emerged (Park et al., 2010).

Chapter 5

Testing Inferred Networks

In this chapter we discuss ways in which inferred networks can be tested for their validity. The tests we consider fall into two broad categories: direct tests of experimentally or computationally inferred networks via further experiments or via confrontation with other data sources—these sources include other networks, expression data, and functional data—and indirect tests of the *predictions* of the network model. An example of an indirect test is the usage of a protein interaction network to predict protein complexes, whose occurrence *in vivo* can be independently verified.

There is necessarily some overlap between the examples discussed here and those discussed in Chapters 3 and 4 on inferring the networks themselves, because it often occurs that the process of validating the predictions of a network model leads to refinement of the network model. This is what happens, for example, when different high-throughput datasets are combined to yield a more accurate network. Furthermore, predictions of one type of network model could well constitute another type of network model. For example, in the previous chapter we outlined how a protein interaction network could be used to predict a genetic interaction network. This prediction task could be viewed either as inference of the genetic network model or, if the predicted genetic network can be independently verified, as a test of the underlying protein interaction network model.

5.1 Co-expression as a Test of Physical Interaction

Perhaps the simplest systematic tests of large protein-protein interaction networks involve testing for co-expression of their corresponding genes or demonstrating that the two proteins are localized in the same intracellular compartment. It is usually not practically feasible to experimentally test co-expression and/or co-localization of every interacting pair of proteins in a protein-protein interaction network. One way around this problem is to examine large datasets of gene expression for evidence of correlated expression across many different conditions.

The assumption here is that gene expression defined in terms of mRNA levels positively correlates with protein expression (or protein abundance). Calculation of Pearson correlation coefficients of normalized expression levels of every pair of genes encoding the corresponding proteins across multiple conditions then estimates the likelihood that their encoded protein products do in fact physically interact. Experimental evidence indeed shows that genes encoding proteins that function in the same multi-protein complexes are often expressed at similar levels (Papp et al., 2003). In fact, as we discussed in Chapter 3, one of the strongest predictors of protein-protein interaction edges appears to be co-expression of their genes. It is therefore only natural that co-expression data are used to validate protein-protein interactions if these data are not used to generate the networks in the first place.

It is important to normalize the expression levels of genes across different experiments because otherwise systematic but spurious high positive or negative correlations might occur among mRNA levels expressed unevenly across experiments. In reality, mRNA expression levels only partially correlate with protein expression levels. It is obviously more useful to test protein-protein interaction networks by examining correlated expression levels of the proteins themselves. However, technical difficulties of measuring thousands of different proteins under hundreds of conditions have so far prohibited implementation of this test in all but the smallest of interaction networks. Improvements in proteomics technologies are expected to significantly impact this area in the future.

Although the above considerations suggest a strong co-expression of genes whose protein products interact, it is important to note that co-expression is not a foolproof test of interaction. A pair of interacting proteins might be co-expressed only under a few special conditions in which the physical contact between the two proteins is biologically significant, while in most other conditions these two proteins might participate in separate or even unrelated functions through their interaction with additional proteins. Thus, a high expression correlation across multiple conditions might potentially provide no information on the existence of physical interaction between the corresponding protein pairs. Conversely, the lack of expression correlation may well convey the false impression that a particular protein-protein interaction edge may be incorrect.

5.2 An Experimental Test of Predicted Protein–Protein Interactions

In reality, there is no single computational test of physical interaction data that is considered a "gold standard"; a variety of methods exist that validate physical interactions. Ultimately, whether or not two proteins do interact physiologically depends on the actual physical demonstration of such an interaction using physical methods such as x-ray crystallography. Here is one important example of a sequence of steps that was taken to test the computational prediction of protein-protein interactions. In this particular example (Markson et al., 2009), the problem was to construct and validate a physical interaction network among a series of important enzymes that have evolved to degrade other proteins.

Protein turnover is an important physiological process. Some proteins are rapidly degraded upon reception of a signal; in the absence of the signal, these proteins are stable, while the rapid decay of the protein in the presence of the signal initiates a specific physiological process. One interesting example is the response of the transcription factor protein p53, which is activated by DNA damage (see Figure 5.1). We remind the reader that this protein was earlier discussed in the context of synthetic lethal interactions in Chapter 2.

When the chromosomal DNA is broken (say, by x-ray radiation or a chemical agent), the broken DNA ends bind to specific proteins and trigger a cascade of enzymatic reactions that leads to the addition of a phosphate group at a particular amino acid of a protein called Mdm2. The unphosphorylated form of this protein normally exists in the cytoplasm and tightly binds to the inactive form of p53. Upon phosphorylation, however, this protein (now called Mdm2p; see Figure 5.1) is rapidly degraded, thus releasing p53 to be transported into the nucleus. Here, p53 is able to regulate a number of target genes that respond to the DNA damage by either promoting DNA repair or by causing programmed cell death. Evidently, suicide is a better option for a multicellular organism than risking the propagation of damaged DNA, which often leads to some form of cancer. The question of interest for us here

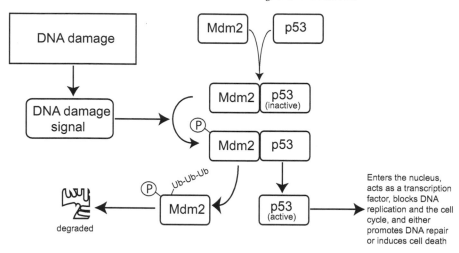

FIGURE 5.1: p53-Mdm2 interaction and DNA damage response. In response to DNA damage, a signaling molecule attaches phosphate groups to the Mdm2 protein, which subsequently dissociates from p53. p53 is then able to enter the nucleus and activate transcription from a number of genes that lead to several different cellular responses: blocking of DNA replication, arrest of the cell cycle, and either promoting DNA repair or triggering cell death. The phosphorylated Mdm2 protein (called Mdm2p) gets tagged by Ubiquitin proteins, and the tagged protein is degraded.

is the mechanism of degradation of the protein Mdm2p. How is it that the protein degradation machinery is so highly selective that it is able to select specifically one protein out of thousands that are present in the cell at any instant of time? It is obviously wasteful for evolution to have designed a special protein-degrading enzyme for every protein. A simple solution happens to be that a relatively small set of enzymes called Ubiquitin E3 ligases can recognize and "tag" larger subsets of enzymes. The tag is a small protein called Ubiquitin. Once a protein is covalently tagged with Ubiquitin, a protein degradation factory called the proteasome degrades that protein. Obviously, one does not expect Ubiquitin E3 ligases to be around in their active forms all the time, otherwise their target protein will be readily destroyed at all times. Activation of a subset of Ubiquitin E3 ligases depends on their interaction with smaller subsets of Ubiquitin ligase activators, which are further activated by other enzymes (Figure 5.2a). Many physiological signals participate in these activation cascades.

To fully understand how the stability of various proteins in the cell is regulated by the Ubiquitin-proteasome system, it is important to know the physical interaction network of ubiquitin ligases with other proteins (Figure 5.2b). How does one proceed with this task?

Because there are numerous E3-target protein interactions to consider, each specific for one E3 enzyme and its target set, it is better to begin with exploring the space of E2-E3 interactions (see Figure 5.2). One possibility is to take a simpler model system in which the E2-E3 interaction space is small, find all E2-E3 interactions by experiments, and then identify the orthologs of each pair in human by computational methods. This is the interolog-based prediction method we encountered in Chapter 3.

Because less than 50% of the E3 proteins in yeast have identifiable orthologues in humans, the techniques for computational identification of the orthologous interactome in humans by inference from yeast two-hybrid data is not possible. For this reason, it was necessary to directly interrogate approximately 1,800 human E2-E3 interaction edges by the yeast two-hybrid technique.

An independent method for validating and predicting E2-E3 interactions was developed using a structure-based homology modeling technique. In this technique, a pair of proteins was predicted to be interacting if they harbored regions that were locally similar to the surface domains of protein pairs known to be interacting (this is very similar to the "interolog" method of Chapter 3). Over 3,000 putative interaction pairs were analyzed using this method. Furthermore, the free energy of binding was also computed for these pairs using methods that are beyond the scope of this book. These free energy scores were binned, and the fraction of pairs in each bin that were detected by the two-hybrid method was calculated. As shown in Figure 5.2c, there is a negative correlation between this fraction and the free energy score: pairs predicted to have low free energy of binding were more likely to be detected in the two-hybrid assay. However, the spread of the data indicated that the free energy score was by itself insufficient for accurate prediction of interaction.

The validity of the two-hybrid interactions was further tested by independent means: pairs of interactors were mutagenized at the predicted interaction surfaces and the mutants were rescanned for two-hybrid interactions. The idea here is that mutations in the predicted interaction surfaces would disrupt the interactions exhibited in a two-hybrid test. Thus, computational prediction of interaction could be tested, at least

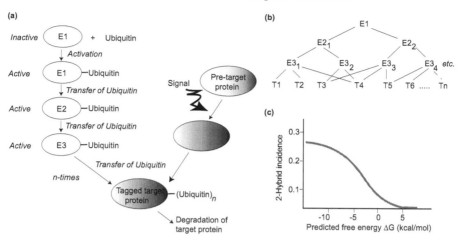

FIGURE 5.2: (a) The reaction cascade for protein degradation by the Ubuquitin tagging system. A few Ubiquitin E1 enzymes are first tagged through autocatalysis by a small protein called Ubiquitin. Upon being tagged, the active E1 enzymes transfer the Ubiquitin protein to a number of different Ubiquitin E2 enzymes, which subsequently transfer the Ubiquitin tag to the Ubiquitin E3 ligases. A specific Ubiquitin E3 ligase then physically interacts with its target protein and transfers the Ubiquitin tag to the target protein. This last reaction occurs several times; when a number of Ubiquitin moieties have been transferred in tandem to the target protein, it is bound by the cell's protein degradation machinery. (b) A diagram of the minimal protein interaction network in the Ubiquitin protein degradation system. The combinatorial interactions between specific subsets of target proteins and a limited number of E3 ligases impart cell- and condition-specific protein degradation. The lines are physical interaction edges. (c) Comparison between computationally predicted and two-hybrid interaction data. Protein homology modeling was used to calculate the free energy of interaction between pairs of proteins, which was binned into groups. The frequency of detection of pairs in each group by a yeast two-hybrid interaction assay is plotted as a function of the predicted free energy of interaction.

for a subset of interactions, by direct experiment. If the predicted surfaces contributed to the two-hybrid scores, it is anticipated that the mutations would degrade the two-hybrid score. Indeed, approximately 75% of the pairs so tested were confirmed to have two-hybrid scores that depended on the degree of sequence conservation of the predicted interaction surfaces.

Ultimately, one of the the best tests for the occurrence of a protein-protein interaction is by a biological function that is dependent on the existence of the interaction. In the above example this was possible because a positive interaction between E2 and E3 proteins should lead to a transfer of Ubiquitin from E2 to E3. Indeed, 93% of the two-hybrid interactions were correlated with biochemical functions predicted for the pairs. However, the absence of a function does not imply the absence of an interaction; it may equally imply that the test could not provide the appropriate conditions for successful function, and may therefore contribute to the false negative rate.

The example of the Ubiquitin ligase protein interaction network serves to illustrate three main points. First, no single method for the detection of protein-protein interactions is definitive. Second, direct experimental detection of interactions coupled with computational prediction wherever possible leads to higher confidence in the interaction network than either method alone. Third, whenever possible, it is important to test the interacting pairs by functional tests.

5.3   Combining Protein Interactions Discovered by High-Throughput Experiments

One of the pressing problems with large-scale screens for protein-protein interactions, such as the two-hybrid screen or co-IP screen, is that the resulting interaction networks, while large, are highly untrustworthy because of error due to both false positives and false negatives. False positives may arise in two-hybrid screens because of nonspecific and unstable "stickiness" of a protein to other proteins (this may in turn be due to the protein having an unusually large number of hydrophobic residues on its surface—see Chapter 7 for a discussion of this). False negatives, on the other hand, may arise due to low coverage of the two-hybrid experiment and due to under-sampling of bait-prey clones within the proteins that are covered. There are also systematic false negatives due to the fact that a bait-prey mating could prove unsuccessful in a two-hybrid experiment, even though the interaction occurs *in vivo* (Huang et al., 2007). The co-IP experiment, while considered a strong test of direct physical interaction, actually only detects participation in the same protein complex. Physical interaction in a co-IP experiment is then inferred by an intervening "matrix" or "spoke" model, as discussed in Chapter 2. Because the "matrix" model includes all prey-prey interactions, it has a high propensity for false positives, while the more conservative "spoke" model may miss out on real prey-prey interactions, leading to false negatives. Further, large scale co-IP experiments are often biased toward the most abundant proteins, leading to loss of coverage and, therefore, to false negatives.

How may these independent, yet error-prone, high-throughput experiments be combined to yield a reliable picture of the true physical interaction network? A solution to this problem must begin with a discussion of a simpler problem first posed and solved by the mathematician George Polya (Polya, 1976).

5.3.1   Polya's Proofreaders

The general problem of combining multiple sets of error-prone data to obtain reliable estimates of error in each dataset (and thus improve the accuracy of the data) was crystallized by Polya into a simple example involving multiple, independent "proofreads" of the same text. Consider a book that has M misprints and is independently read by two proofreaders A and B. A notices m_1 misprints in the book, B notices m_2 misprints, and m_3 misprints are noticed by both. Note that the misprints counted in m_3 are also counted in m_1 and m_2, and therefore the total number of detected misprints is $m_1 + m_2 - m_3$. The actual number of misprints, M, is unknown and so are the propensities for A and B to detect errors accurately. Let us represent these propensities as follows. Let p_A be the probability that A will detect a true misprint, and let p_B be the analogous probability for B. Then the mean number of misprints detected by A will be Mp_A, the mean number of misprints detected by B will be Mp_B, and the mean number of misprints detected by both will be $Mp_A p_B$ (the probabilities p_A and p_B are multiplied because A and B are assumed to be independent readers). When the numbers of misprints (true as well as detected) are large, the detected numbers of misprints will be well approximated by the mean numbers. Therefore,

$$m_1 = Mp_A, \quad m_2 = Mp_B, \quad m_3 = Mp_A p_B. \tag{5.1}$$

Solving these three equations for the three unknowns M, p_A, and p_B yields

$$M = \frac{m_1 m_2}{m_3}, \quad p_A = \frac{m_3}{m_2}, \quad p_B = \frac{m_3}{m_1}. \tag{5.2}$$

This elegant method therefore enables one to estimate not just the error rates for the two proofreaders, but also the number of misprints that are missed by both of them. This number is the true number of misprints minus the number of detected misprints, which is $M - (m_1 + m_2 - m_3) = (m_1 - m_3)(m_2 - m_3)/m_3$.

We now discuss how errors in high-throughput datasets and the true number of protein-protein interactions can be similarly estimated and how these estimates can be used to assign confidence values to individual interactions. The interactions with highest confidence can then be assessed for their biological significance.

5.3.2 Integration of Independent High-Throughput Datasets

Polya's proofreading problem has a direct analogy in the world of protein interaction data. The "misprints" here correspond to protein-protein interactions, and the independent "proofreaders" are the independent high-throughput experiments. However, the protein interaction problem is somewhat complicated by the fact that the experiments can have false positive errors (as described above) in addition to false negative ones. If we use data from two high-throughput experiments, that leads to two additional parameters to estimate. In this case, three equations no longer suffice to estimate all unknown parameters. Therefore, we need data from at least three independent experiments to estimate all unknown parameters: the false positive and false negative error rates in each experiment in addition to the total number of true protein-protein interactions in the network. The situation can be somewhat simplified, and the estimates made more accurate, if one of the three experiments is a "gold-standard" experiment, that is, it has negligible false positive rate. The measured data consist of the number of interactions found in each experiment individually (three numbers), the number of common interactions found in two of three experiments (three numbers), and the number of common interactions found in all three experiments, resulting in a total of seven observations. The unknown parameters are false positive rates in two experiments, false negative rates in all three experiments, and the total number of true interactions, resulting in a total of six parameters.

The above framework was outlined in work by d'Haeseleer and Church (2004), who used a "gold-standard" reference set of 1,542 yeast protein-protein interactions culled from the MIPS database. The other two datasets were chosen from a collection of yeast two-hybrid and co-IP experiments. Gold-standard datasets, because they are more accurate on the interactions that they cover, typically have low coverage of the network and therefore do not report many existing interactions.

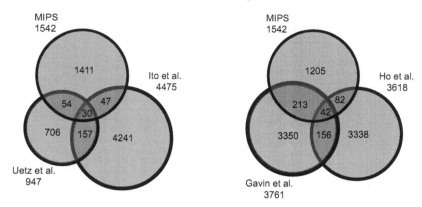

FIGURE 5.3: Relative sizes and intersections of a gold-standard MIPS reference dataset of protein-protein interactions and 2 two-hybrid experiments (Uetz et al., 2000; Ito et al., 2001) (left panel), and two co-IP experiments (Gavin et al., 2002; Ho et al., 2002) (right panel). The numbers directly below the dataset labels refer to the total number of interactions in each experiment, while the numbers within the circles refer to the number of interactions in each intersection and its complement. While there are ten numbers in each panel, any three of them can be expressed in terms of the others, resulting in seven independent counts.

Another way of saying this is that they have a high negative false negative rate. The overlap of high-throughput datasets that have good coverage (i.e., low false negative rate) but high false positive rate with gold-standard datasets that have low false positive rate but less coverage is a common technique employed in network biology for estimating errors in interaction data, and for ultimately discarding interactions that are highly error-prone.

Figure 5.3 displays the intersections (d'Haeseleer and Church, 2004) of the gold-standard MIPS reference set, in one case, with two yeast two-hybrid experiments, and in the other case, with two co-IP experiments. With three datasets chosen at a time in this manner, and with one of them always being the fixed MIPS reference set, we can write the equations that will determine various unknown parameters. Table 5.1 lists the observed and unknown variables in the problem. Note that the false positive rates for high-throughput experiments 1 and 2 are $1 - q_1$ and $1 - q_2$, respectively, and the false negative rates are $1 - p_1$ and $1 - p_2$, respectively. An additional known count in the problem is the total number of unique protein pairs, which we call P. Thus, $P - N$ is the total number of true noninteractions.

TABLE 5.1: **Variables Related to the Estimation of Error Rates in High-Throughput (HTP) Protein Interaction Experiments**

Type	Description	Symbol
	No. of interactions in ref. dataset	g
	No. of interactions in HTP experiment 1	n_1
	No. of interactions in HTP experiment 2	n_2
Observed	No. of interactions in $1 \cap 2$	n_{12}
	No. of interactions in ref. $\cap$ 1	g_1
	No. of interactions in ref. $\cap$ 2	g_2
	No. of interactions in ref. $\cap$ 1 $\cap$ 2	g_{12}
	Pr(true interaction correctly reported by ref.)	p
	Pr(true interaction reported by 1)	p_1
	Pr(true interaction reported by 2)	p_2
Unknown	Pr(true noninteraction reported by 1)	q_1
	Pr(true noninteraction reported by 2)	q_2
	True number of total interactions	N

By analogy to Polya's proofreaders, we can equate each observed count to an expected count. However, because of the presence of false positives, each observed count (except the counts involving the reference set) gets two contributions: one from true interactions being correctly reported, and one from true noninteractions being falsely reported as interactions. Therefore, the relevant equations are

$$g = Np$$
$$n_1 = Np_1 + (P - N)(1 - q_1)$$
$$n_2 = Np_2 + (P - N)(1 - q_2)$$
$$n_{12} = Np_1p_2 + (P - N)(1 - q_1)(1 - q_2)$$
$$g_1 = Np_1p$$
$$g_2 = Np_2p. \tag{5.3}$$

The above constitute six equations that may be solved for the six unknown parameters. In addition, there is an equation for g_{12} as a product of N, p, p_1, and p_2, but in the case when the reference network has no false positive error and the datasets are independent, this additional equation is unnecessary and serves as a consistency check on the observed data.

In reality, however, the assumptions that the reference dataset is without false positive error, and that the three datasets are independent, are never quite true. Indeed, a quick calculation using the counts in Figure 5.3 reveals that the counts are not consistent with these assumptions (the reader may try this as an exercise!). How do we then infer the error rates and the true total number of interactions? We will try to answer this question somewhat qualitatively here.

First, let us suppose that the datasets can be treated as independent but the MIPS reference dataset actually has some false positive error. It is trivial to model this error in a manner identical to that for the two HTP datasets. The false positive error of the reference dataset then becomes an additional unknown parameter and modifies the first equation and the last two equations in Equation (5.3) above. Because of the additional unknown parameter, we need an additional equation— the equation for g_{12}—which is no longer redundant.

Violation of the independence assumption is more serious, and that can happen in various ways. Consider, for example, a situation where the HTP datasets are pairwise independent but the reference dataset is not independent of the HTP datasets. In this case, the last two equations in Equation (5.3) are changed to $g_1 = Np_1p_{|1}$ and $g_2 = Np_2p_{|2}$, where $p_{|1}$ is the probability that a true interaction is correctly reported by the reference dataset *given* that it was correctly reported by experiment 1, and similarly for $p_{|2}$. This introduces two new parameters in general, but only one new parameter if we demand that $p_{|1} = p_{|2}$, which is tantamount to requiring that the probability for the reference set to report an interaction correctly is biased *in the same way* irrespective of whether the interaction was also reported by either HTP dataset 1 or HTP dataset 2. For this requirement to be satisfied, it is necessary that both HTP datasets, although independent, should reflect the same *type* of experiment. Therefore, for example, both HTP datasets in the intersection should either correspond to two-hybrid experiments or to co-IP experiments, as indicated in Figure 5.3. With this proviso in place, d'Haeseleer and Church found that the error rates inferred in this manner for HTP experiments were quite insensitive to which additional HTP experiment was used in the intersection.

Using these methods, d'Haeseleer and Church, and later Hart et al. (2006), estimated error rates of various HTP datasets as well as the size of the yeast interactome. The false positive rates for HTP data range from about 45% to as much as 90%, while the estimated size of the yeast interactome is roughly 50,000 interactions. Further, after error rates have been computed, it is possible to re-examine each pair of proteins

(A and B below) and assess, in Bayesian fashion, the probability that they truly interact *given* the results of various HTP experiments:

$$\Pr(\text{A and B interact} \mid \text{expt.}_1, \text{expt.}_2, \dots) \propto$$

$$\Pr(\text{expt.}_1, \text{expt.}_2, \dots \mid \text{A and B interact})\Pr(\text{A and B interact}), \quad (5.4)$$

and a similar formula holds for the probability of a true noninteraction. The proportionality factor in the above equation can be found by requiring that the probabilities for interaction and noninteraction sum to 1. On the right-hand-side of the above equation, $\Pr(\text{expt.}_1, \text{expt.}_2, \dots \mid$ A and B interact) can be found using the inferred false positive rates and dependency assumptions among experiments, while $\Pr(\text{A and B}$ interact) can be found using the inferred total number of true interactions relative to the total number of protein pairs. Thus, the judicious combination of high-throughput interaction datasets can be used to derive a measure of confidence for every interaction and noninteraction. It is therefore possible to rank the interaction pairs according to their confidence levels, which in turn can be used to ascribe biological functions to the high-confidence pairs according to the method described below.

5.4 From Curated Protein–Protein Interactions to Protein Complexes

We have discussed how distinct error-prone protein interaction datasets may be combined to yield estimates of the true probability of interaction between two proteins. The resulting protein interaction network may then be represented as a *weighted graph*, that is, a network in which an edge (interaction) between two nodes (proteins) has a numerical weight corresponding to the probability of that interaction. One of the crucial tests of the accuracy of such a weighted graph is its ability to faithfully reconstruct biological function. Many cellular processes are performed by stable protein *complexes*: sets of interacting proteins that carry out their functions as a single unit. Therefore, an indirect test of the accuracy of a protein interaction network is the validation of the protein complexes that are predicted by the network.

Unfortunately, however, the pattern of protein complexes that are consistent with a given protein interaction network is not unique. Rather, it depends on the method used to reconstruct the complexes from the network. Complex identification requires merging of appropriate sets of protein interactions, an additional inference step beyond the identification of individual protein interactions. In principle, if one method was definitively known as the "best" method to identify protein complexes from individual interactions, then the protein complexes predicted by that method could be compared to known protein complexes in order to test the accuracy of the protein interaction network itself.

Like many problems in network biology, however, the problem of complex identification from protein interaction data is an active area of research, and no one method can be conclusively recommended. Thus, while we are not in a position to unambiguously test the accuracy of a protein interaction network using predicted complexes, it is safe to conclude that if many well-validated computational methods correctly and robustly identify the same set of complexes, then the underlying protein interaction network is probably trustworthy. On the other hand, if there exists large variation among the predictions from different methods, then the network itself is probably suspect.

Here we discuss two of the many methods used to identify complexes from curated protein interaction data. One method is largely *unsupervised*, that is, it does not require prior knowledge of the specific nature of specific protein complexes, while the other method is *supervised*, that is, it predicts new complexes by learning from the properties of known complexes.

5.4.1   Complex Identification Using Markov Clustering

Most unsupervised methods to identify complexes in protein interaction networks employ the general idea that complexes correspond to locally dense *clusters* in the network; they are characterized by a large proportion of within-cluster interactions and a small proportion of interactions across clusters. Thus, unsupervised methods to identify complexes are essentially methods that identify clusters or *modules*. The Markov clustering method devised by van Dongen (2000) is particularly useful because of the ability to tune the "granularity" or compactness of the identified clusters.

Because clusters are locally dense regions in a network, one expects that the number of paths of a given length between two nodes in the same cluster should be high relative to the number of such paths between two nodes in different clusters. Thus, a "random walker" on the network will frequently take steps within the same cluster and only infrequently transition across clusters. Now suppose there is a one-step transition probability matrix **T** that determines the movement of the random walker on the graph. Its elements T_{ij} correspond to the probability that a random walker who starts at node j of the graph reaches node i in one step. Such a matrix is also called a *stochastic matrix* on the network. For a weighted graph, a stochastic matrix can be naturally constructed by appropriately normalizing the edge weights. For example, consider the network of protein-protein interactions whose edge weights w_{ij} correspond to probabilities of interaction. We can construct a stochastic matrix for this network by defining $T_{ij} = w_{ij} / \left(\sum_k w_{kj} \right)$. The random walker will therefore have higher probability to transition across edges that have stronger evidence for protein-protein interaction.

There are two key operations that lie at the heart of the Markov cluster approach. First, note that a matrix power of a stochastic matrix is also a stochastic matrix corresponding to a many-step transition probability. For example, $\mathbf{T}^2$ is a stochastic matrix whose elements correspond to probabilities of transition from one node to another in two steps. The operation of taking the matrix power of a stochastic matrix is called *expansion*, because it corresponds to computing random walks of higher length. Because higher length paths are more prevalent within clusters than between clusters, the expansion operation associates higher probabilities to node pairs within clusters. At the same time, the new stochastic matrix resulting from expansion is less sparse than the old one, as nodes that had no direct "one-step" connections may still be connected by multiple steps.

A second operation, called *inflation*, consists of taking powers T_{ij}^r of each element of a stochastic matrix and then renormalizing the matrix to obtain a new stochastic matrix. Mathematically, this operation has the effect of increasing the probability of more reliable edges while decreasing the probability of less reliable ones. Thus, inflation serves to concentrate the stochastic matrix on reliable interactions, while expansion causes the paths of a random walker on the network to become more evenly spread. If expansion and inflation are applied in an alternating fashion to the stochastic matrix, the resulting graph becomes more and more clustered. These clusters are then predicted to be protein complexes. The final prediction can be somewhat varied depending on the values of the power exponents in the expansion and inflation operators, which is what lend flexibility to the method.

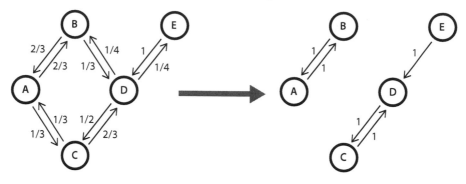

FIGURE 5.4: Markov clustering applied to a small network of five nodes. A connected network of five nodes, labeled A–E, is shown in the left panel. Numbers signify transition probabilities associated with the transitions indicated by arrows. These transition probabilities can be defined even for an undirected weighted network (such as a protein interaction network) as described in the text. The right panel shows the resultant network after five iterations of alternating expansion and inflation. In this example, the power parameters (denoted by r in the text) for the inflation and expansion operations are 5 and 3, respectively. As shown, all transition probabilities converge to 1 or 0 in a matter of five iterations, and two clusters are readily identified. A similar procedure applied to large protein interaction networks may be used to identify protein complexes.

The Markov clustering approach as applied to a small network is illustrated in Figure 5.4 and Box 5.1.

Krogan et al. (2006) used the Markov clustering approach to identify complexes in a weighted yeast protein interaction network generated by a combination of experimental and machine learning approaches. They identified 547 nonoverlapping protein complexes and analyzed (1) the agreement of these complexes with known yeast complexes, (2) homogeneity of biological function of proteins within the same predicted complex, and (3) co-localization within the cell of proteins within the same predicted complex. In all cases, they reported better results than previous approaches (better agreement with known complexes, more functional similarity within complexes, and less variation in cellular location among proteins within a complex), a finding attributed to the accuracy of the protein interaction network that they constructed. The complex identification process thus led ultimately to a validation of the underlying network. As stated earlier, such validation is clearly not foolproof, as the inability to find validated complexes could well be a shortcoming of the complex identification algorithm rather than the un-

Box 5.1 Matrix operations in Markov clustering: an example.

The two fundamental operations in Markov clustering, namely expansion and inflation, are illustrated here with an example. Consider the five-node network shown in the left panel of Figure 5.4, where probabilities of transition between nodes are shown next to the arrows that signify each possible transition. The stochastic matrix of this network is

$$\mathbf{T} = \begin{pmatrix} 0 & 2/3 & 1/3 & 0 & 0 \\ 2/3 & 0 & 0 & 1/3 & 0 \\ 1/3 & 0 & 0 & 2/3 & 0 \\ 0 & 1/4 & 1/2 & 0 & 1/4 \\ 0 & 0 & 0 & 1 & 0 \end{pmatrix},$$

where rows and columns of $\mathbf{T}$ are ordered according to the alphabetical order of the nodes. As stated in the text, the *expansion* operation consists of taking a power of the matrix. Let us assume for simplicity that the power parameter $r = 2$. Then each expansion step consists of squaring the stochastic matrix:

$$\mathbf{T}^2 = \begin{pmatrix} 0 & 2/3 & 1/3 & 0 & 0 \\ 2/3 & 0 & 0 & 1/3 & 0 \\ 1/3 & 0 & 0 & 2/3 & 0 \\ 0 & 1/4 & 1/2 & 0 & 1/4 \\ 0 & 0 & 0 & 1 & 0 \end{pmatrix}^2 = \begin{pmatrix} 5/9 & 0 & 0 & 4/9 & 0 \\ 0 & 19/36 & 7/18 & 0 & 1/12 \\ 0 & 7/18 & 4/9 & 0 & 1/6 \\ 1/3 & 0 & 0 & 2/3 & 0 \\ 0 & 1/4 & 1/2 & 0 & 1/4 \end{pmatrix},$$

while an inflation step consists of squaring each element of $\mathbf{T}$ followed by normalizing the rows so that each row sums to 1:

$$\begin{pmatrix} 0 & 4/9 & 1/9 & 0 & 0 \\ 4/9 & 0 & 0 & 1/9 & 0 \\ 1/9 & 0 & 0 & 4/9 & 0 \\ 0 & 1/16 & 1/4 & 0 & 1/16 \\ 0 & 0 & 0 & 1 & 0 \end{pmatrix} \xrightarrow[\text{(divide by row sum)}]{\text{normalize}} \begin{pmatrix} 0 & 4/5 & 1/5 & 0 & 0 \\ 4/5 & 0 & 0 & 1/5 & 0 \\ 1/5 & 0 & 0 & 4/5 & 0 \\ 0 & 1/6 & 2/3 & 0 & 1/6 \\ 0 & 0 & 0 & 1 & 0 \end{pmatrix}.$$

One observes that an expansion step causes the spreading of probability: there are fewer zeros in the $\mathbf{T}^2$ transition matrix than in the $\mathbf{T}$ transition matrix, while an inflation step causes greater concentration in high-probability transitions (the ratio of the A → B transition probability to the A → C transition probability increases from 2 to 4 after one inflation step).

In general, the power parameter r can be arbitrary. For example, the network shown in the right panel of Figure 5.4 is achieved using $r = 5$ for inflation and $r = 3$ for expansion and after five alternating iterations of expansion and inflation.

derlying network. However, it is equally clear that inaccurate networks will lead to inaccurate prediction of complexes, no matter how good the complex identification procedure is. It is therefore useful to employ more than one computational method for complex identification, as we have discussed.

5.4.2 Complex Identification Using Supervised Clustering

One of the drawbacks of Markov clustering, and of unsupervised methods in general, is that it tacitly makes assumptions about the types of clusters one is looking for. Specifically, the Markov clustering method assumes that protein complexes form "clique-like" structures in the network. Cliques are sub-networks that are densely connected within themselves but have few connections with the rest of the network. While the assumption of "cliquishness" of complexes seems quite reasonable and, in fact, holds for most complexes, complexes that do not have this property may be missed by the method (false negatives) and, on the other hand, cliques in the protein interaction network that do not correspond to true functional complexes may be identified as such (false positives). It is therefore reasonable to consider alternative methods for complex identification that do not make simplifying *a priori* assumptions about the topology of protein complexes, but instead learn from the structure of known complexes.

A simple yet powerful supervised approach for complex identification is based on a Bayesian network model (Qi et al., 2008). We have previously encountered dynamic Bayesian networks for predicting regulatory interactions. Here, the problem can be stated as follows. Given a sub-network of a protein interaction network, we wish to find the probability that this sub-network represents a protein complex. To model this probability, we first identify several features of the sub-network that are expected to be distributed differently among true protein complexes than among "non-complexes." Let us call these features $F_1, F_2, \ldots, F_m$. One such feature that immediately comes to mind is the size (i.e., number of proteins) in the sub-network; it is reasonable to assume that the size distribution of complexes would be different from the size distribution of non-complexes. We call this feature n, the number of nodes in the sub-network. The desired probability that the sub-network is a true protein complex can then be written as the posterior probability

$$P(c = 1 \mid n, F_1, \ldots, F_m) \propto P(n, F_1, \ldots, F_m \mid c = 1)P(c = 1), \quad (5.5)$$

where c is the class label for the sub-network, with $c = 1$ indicating that the subgraph is a complex and $c = 0$ indicating that the subgraph is not a complex. The above equation follows from Bayes' rule.

How do we compute the probabilities on the right-hand side of Equation (5.5)? Because this is a supervised method, the answer lies in using data from known protein complexes. We can calculate n and the remaining features for a set of known protein complexes and thus, at least in principle, empirically estimate the joint distribution of these features for known complexes, which is the first factor on the right-hand side of Equation (5.5). Empirical estimation of such a highly multivariate distribution is, however, unreliable unless one has access to a very large amount of data, and it is best to proceed by simplifying the multivariate distribution in some way. Such a simplification is possible by constructing a Bayesian network model that elaborates the dependencies among the various features. Qi et al. (2008) assumed the Bayesian network model of Figure 5.5, in which the features $F_1, F_2, \ldots, F_m$ are conditionally independent given the subgraph size and the class label of the subgraph. Under this assumption, the joint probability of Equation (5.5) factorizes as follows:

$$P(n, F_1, \ldots, F_m \mid c = 1) = P(F_1, \ldots, F_m \mid n, c = 1)P(n \mid c = 1)$$

$$= \prod_{i=1}^{n} P(F_i \mid n, c = 1) \, P(n \mid c = 1). \quad (5.6)$$

Putting Equations (5.5) and (5.6) together yields the *posterior odds ratio* for a candidate sub-network:

$$\frac{P(c = 1 \mid n, F_1, \ldots, F_m)}{P(c = 0 \mid n, F_1, \ldots, F_m)} = \frac{P(c = 1)P(n \mid c = 1) \prod_{i=1}^{n} P(F_i \mid n, c = 1)}{P(c = 0)P(n \mid c = 0) \prod_{i=1}^{n} P(F_i \mid n, c = 0)}. \quad (5.7)$$

The various probabilities on the right-hand side of the above equation are straightforward to estimate from known complexes. We can estimate the ratio of numbers of complexes to non-complexes in protein interaction networks from known complex data [this yields $P(c = 1)/P(c = 0)$], we can estimate the individual distributions of each feature in known complexes and non-complexes [this yields $P(F_i \mid n, c = 1)$ and $P(F_i \mid n, c = 0)$ for each feature], and we can estimate the size distributions of complexes and non-complexes [this yields $P(n \mid c = 1)$ and $P(n \mid c = 0)$]. Candidate sub-networks with a posterior odds ratio greater than 1 can then be identified as complexes in this method.

The features $F_1, F_2, \ldots, F_m$ used by Qi et al. to discriminate between complexes and non-complexes are mainly topological features of

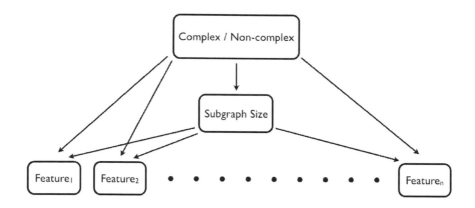

FIGURE 5.5: The Bayesian network model architecture of Qi et al. (2008). In this model, it is assumed that the distributions of the features $F_1, F_2, \ldots, F_n$ are determined by whether or not the subgraph is a protein complex and by the size of the subgraph. Furthermore, given these two pieces of information, the features are independent. Finally, the model assumes that the distribution of subgraph size is determined solely by whether or not the subgraph is a protein complex.

the candidate sub-network. They include properties that we examine in Chapter 7: the density, clustering coefficient, features of the degree distribution of the nodes in the sub-network, and so on. However, the Bayesian network framework itself is quite generic and there is no reason why other features, including biological features (e.g., properties of the mRNA/protein expression distribution among the nodes of the subgraph), cannot be incorporated into the Bayesian network.

Qi et al. estimated distributions of features among known complexes in the MIPS database. They estimated feature distributions for non-complexes by selecting nodes at random from the protein-protein interaction network. They then developed a simulated annealing method (which is a computational optimization method that is beyond the scope of this book) to search for sub-networks that have a high posterior odds ratio (i.e., the putative protein complexes). Finally, they measured the success of their method by finding the overlap between newly identified complexes and existing ones. Their results significantly outperformed those from other unsupervised and supervised methods.

5.5 Testing DNA–Protein Interactions

We now turn to the testing of physical interactions between DNA and protein. For such interactions to be biologically meaningful, the binding of the protein to the specific DNA sequence should constitute a detectable biological function.

Note that here we consider only protein-DNA interactions that are specific for particular DNA sequences. Thus we will not consider non-specific interactions that are independent of DNA sequence. We will also not consider cases where the sequence preference is so ubiquitous that thousands of interactions exist for the protein in question. Examples of this scenario include interactions between histone proteins and DNA, and those between methylated DNA-binding proteins and methylated DNA.

In Chapter 3, we used gene expression as a means for computationally predicting DNA-protein interaction. The problem we address here is how these computational predictions may be suitably tested.

Highly sequence-specific protein-DNA interaction edges usually imply that the protein is a transcription factor, which could be negatively or positively acting, or, very rarely, a sequence-specific DNA modifying enzyme (e.g., one that may cleave the DNA at a particular sequence motif). If the protein is known to be a transcription factor, then a test of a specific interaction edge might constitute examining the prediction that if the protein is expressed, then its target gene is either expressed or not expressed at a high level—that is, determining the existence of a correlation that is high in magnitude (i.e., highly positive or negative) between the expression levels of the protein and its target gene's mRNA across many different conditions. This is a static inference, in which we make no obligatory reference to the timing of expression of the protein and its target gene, although the difference in expression timing might itself constitute a "condition." A test such as this may take the form of comparing two networks, a protein-DNA interaction network and a network of co-expression.

A more precise test of a protein-DNA interaction may involve testing a dynamic property of the interaction. For example, if a protein is a positive (or negative) regulatory factor of its target gene, then we predict that a change in the expression level of the protein (or its corresponding mRNA) with time should positively (or negatively) correlate with changes in the mRNA level of its target gene. This in principle should be measurable in a time-series experiment, in which RNA samples are collected across multiple time points during which the gene expression levels are not in equilibrium, and the time variation of the mRNA level corresponding to the DNA binding protein should systematically predict the time variation of its target gene's mRNA level. Note that this is exactly the type of correlation that was exploited in the REVEAL and related methods (discussed in Chapter 3) to *predict* protein-DNA interactions in the first place. It is clear, therefore, that expression data can only be used to test experimentally discovered DNA-protein interactions, not computational predictions.

An interesting way of testing a protein-DNA interaction edge involves testing a genetic prediction. For example, mutation of a specific DNA sequence at the target site of the protein is expected to inhibit the binding of the protein to the DNA target, and thus should cause a change in the expression level of the target gene. The converse experiment can also be carried out: mutating the gene that encodes the protein should have the effect of changing expression levels of its target genes under at least one given condition.

If none of the above effects can be demonstrated, then it may mean that the protein-DNA interaction edge does not exist, that the interaction edge has no biological significance, or that we have not yet discovered the exact conditions under which the function actually takes place.

A marvelous example of how complex DNA-protein interaction networks are built and tested is provided by work in C. Murre's laboratory (Lin et al., 2010). B-cells are lymphocytes that participate in immunity against invading organisms specifically by rearranging their genes for making antibodies against foreign proteins (see Box 3.4). They are formed from a collection of somatic (i.e., adult) stem cells that produce all blood cells. It was earlier known that three critical transcription factor proteins—E2A, EBF1, and FOXO—are important for converting precursor cells into B-cells by selectively regulating hundreds of genes. What genes do these transcription factors regulate, and what is the network of interacting proteins and DNA that is important for making B-cells out of their precursor cells?

To address this question, Murre's laboratory identified the DNA binding sites of the three transcription factors by directly measuring the occupancies of these proteins on the genome at three different stages during the development of the precursor B-cells to the mature B-cells.

Interestingly, the extent of occupancy of these proteins changed during development. For example, E2A bound more than twice as many sites at a later time in development as it did at an earlier time, illustrating the dynamic nature of DNA-protein interaction networks.

What controls which DNA sites are occupied at a particular time by the same protein? The answer seems to lie in a complex code influenced by the other DNA binding proteins bound at those sites. As just one example, a particular modified form of a histone protein (H3K4-methylated) that wraps around the DNA without much sequence specificity was found to co-locate with E2A.

To generate a combined network of protein-DNA interaction related to B-cell development (Figure 5.6), Murre's laboratory joined hands with Ideker's laboratory (Lin et al., 2010). First they identified genes that showed differential expression during B-cell development, reasoning that the three transcription factors should directly regulate at least some of these genes. To remove most of the indirectly regulated genes from this set, they subsequently filtered out those genes exhibiting differential regulation during B-cell formation but whose protein products were not situated next to observed DNA occupancy sites of the three transcription factors at the corresponding stages. This yielded nearly 2,000 target genes for the three transcription factor proteins. As expected for the network of genes regulated by these transcription factors, the target gene set was highly statistically enriched for gene functions involved in lymphocyte development. This statistical enrichment of the gene set for lymphocyte-development-related functions is considered a validation of the initial observation of transcription factor binding to the DNA target genes.

5.6 Congruence of Biological Function among Interacting Partners

An important aspect of the process of testing interactions is to classify genes according to function so that the functional congruence between interacting genes/proteins can be assessed. This assessment is facilitated by the Gene Ontology (GO) procedure [http://www.geneontology.org; Ashburner et al. (2000)] that systematically classifies gene and protein function.

Ontology is the study of the relationships of entities, such as genes

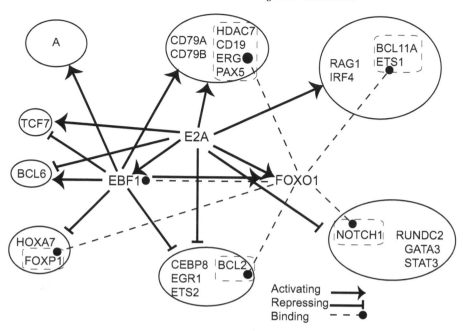

FIGURE 5.6: The network of protein-DNA interaction in B-cell development. The three central transcription factors—E2A, EBF1, and FOXO1—bind to target DNA sequences nearby approximately 2,000 genes. The expression levels of these genes are regulated during B-cell development. The target gene sets are diagrammatically represented as ovals, with the size of an oval increasing with the number of genes in the corresponding set. The arrows show activating protein-DNA interactions, and bars represent repressive interactions. Dotted lines ending in filled circles represent FOXO1 binding to the DNA near the regulatory regions of genes in the sets. A few representative genes that are known to play important roles in B-cell development in each set of target genes are separately shown in dashed boxes. A few other important genes in some of the sets are also noted. The gene set "A" did not appear to have any known B-cell development related genes when this network was reported. Adapted and simplified from Lin et al. (2010).

or their encoded proteins, with other entities (e.g., other genes or proteins), and the grouping of such entities based on their common or shared properties. The Gene Ontology (GO) Consortium was established in 1998 as a collaborative effort of biologists studying a few model organisms. The original aim of the consortium was to address the need for a uniform description of gene products and their relations across different databases. GO has now matured into a general description and classification system for genes and gene products across entire biological systems, including humans.

GO presents a structured and controlled vocabulary to describe gene products in terms of three ontological descriptor classes: their associated *biological processes*, their *cellular components*, and their *molecular functions*. These descriptors are independent of the species of organisms, and are arranged in the form of three separate acyclic directed graphs.

Near the root of each of the three ontologies are a few unique ontological descriptions, and from these general ontologies one can traverse along the acyclic graphs through the branches to the leaves—thus encountering more and more detailed ontological descriptions. For example, at the highest level of the *biological process*, one of the ontological descriptors is "DNA Metabolism" (Figure 5.7). One level down are several finer ontological subgroups, such as "DNA Replication," "DNA Repair," "DNA Recombination," etc. Below this level, there are further specialized ontological descriptors, such as "DNA Ligation," "DNA Unwinding," etc. A gene called CDC9, which encodes a protein that has the biochemical property of ligating (i.e., joining) two pieces of DNA, is a leaf in this tree and is classified by the ontological property of "DNA Ligation." However, this gene's protein product has functions corresponding to two separate biological processes: "DNA Repair" as well as "DNA Recombination." Hence, CDC9 is reachable from both of these branches, as shown in Figure 5.7. Furthermore, the "DNA Replication" process has a subprocess called "DNA-dependent DNA Replication," which also reaches "DNA Ligation." Thus, the ontological description of CDC9 is reachable from "DNA Metabolism" along three different branches of the GO hierarchy. The reader should note, however, that there may be multiple genes annotated with exactly the same ontology. For example, both CDC9 and LIG4 genes have the same GO annotation but they are two distinct genes.

Similarly, the *molecular function* ontological description has a hierarchical tree associated with it, which classifies in finer and finer ways the function of a gene product to molecular detail. Figure 5.8 shows an actual example of the results of GO analysis carried out on a set of genes. The classification tree shows that some genes are annotated with the molecular function termed "DNA Binding," which are further classified into two groups: "Sequence-specific DNA Binding" and "Transcription Factor Activity." The latter genes are also annotated with a higher-level descriptor, that of a "Transcriptional Regulator." Thus, for example, a

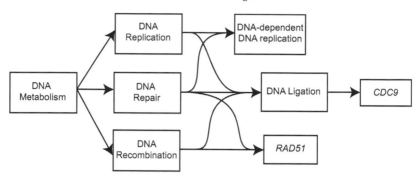

FIGURE 5.7: An example involving the "DNA Metabolism" Biological Process and its sub-processes in the Gene Ontology hierarchy.

"Transcriptional Regulator" need not be a "Transcription Factor" but the latter is always a "Transcriptional Regulator."

The third ontology, the *cellular component* category, relies on the idea that the basic structural organization of all cellular life is the same or is based on very similar basic motifs. Examples include the nucleus, cytoplasm, mitochondria, plasma membrane, and so on. These components also display hierarchical organization. For example, the nuclear matrix is a subcompartment of the nucleus, which itself is a subcompartment of the cell. In an actual example of GO cellular component analysis (Figure 5.9), a group of genes is classified into two separate groups on the basis of the physical organization in the cell of their protein products: "Proteinaceous Extracellular Matrix" and "Neurofilament."

The common terminology presented by GO for the purpose of defining genes and their products and for uniformly classifying the functions of these entities leads to consistent and unambiguous understanding of the contextual biological significance of genes. Such a database also allows the research community to assign properties of genes or gene products at different levels of "granularity."

A major effort in molecular biology research is to infer the biological functions of genes from co-expression data. The most useful ontology for relating biological function to co-expression is the Biological Process, given the *ad hoc* but convenient assumption that a pair of co-expressed genes function in the same or related biological pathway(s). However, molecular functions of these two genes might be distinctly different, or their products might be localized in different intracellular compartments, such that their respective positions in the GO Molecular

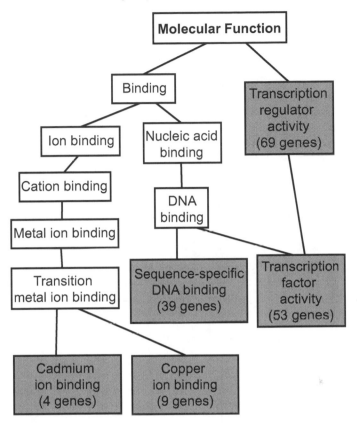

FIGURE 5.8: An example in which multiple genes share the same Molecular Function category (i.e., binding) in the Gene Ontology hierarchy. The relationships shown were obtained by querying the GO database [http://www.geneontology.org; Ashburner et al. (2000).]

Function or Cellular Component schema might indeed be different.

5.7   Testing Genetic Interactions

As we know, genetic interactions can have varied interpretations depending on their type. For example, an allele-specific suppression interaction implies a protein-protein physical interaction (see Chapter 1). Synthetic lethal interactions, on the other hand, have more complex consequences for physical interactions.

One interpretation of synthetic lethality is that members of synthetic

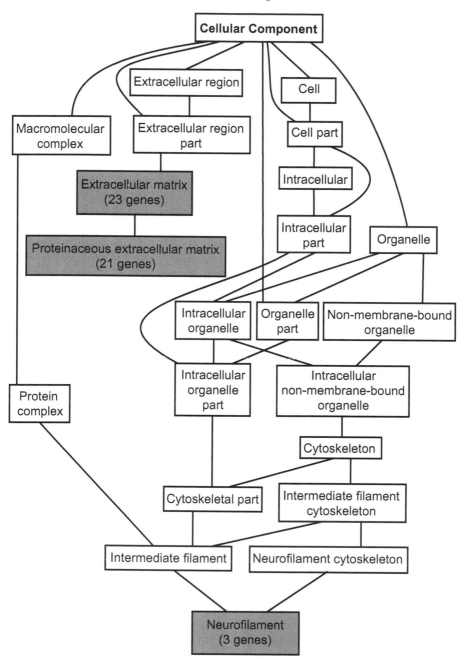

FIGURE 5.9: An example illustrating the Cellular Component category in the Gene Ontology hierarchy. The resolution of protein location increases as one reads down the heirarchy. The relationships shown were obtained by querying the GO database [http://www.geneontology.org; Ashburner et al. (2000).]

lethal pairs function in parallel but redundant pathways, such that the loss of any two members of the same pathway causes no appreciable difference to the organism but the loss of a member each from the two redundant pathways is lethal (see Figure 1.5). Indeed, classification of most synthetic lethal genes that encode known proteins belonging to well-established biochemical pathways indicates that the majority of synthetic lethal pairs map across different pathways rather than within the same pathway (Kelley and Ideker, 2005). Thus, they mostly follow the patterns of Figure 1.5a and Figure 1.5b. We therefore expect that GO classification of synthetic lethal pairs should reveal systematic correlation in the biological functions of the two genes in a pair. In other words, pairs of synthetic lethal genes should have short Biological Process GO distances separating them. This is because when two separate pathways that converge to the same essential biological process are simultaneously disrupted, synthetic lethality occurs.

On the other hand, the two members of the synthetic lethal gene pair need not be related in their Molecular Function or Cellular Component GO ontologies. This is because two interacting proteins that participate in the same biological process may have different biochemical functions. For example, Mdm2p is a Ubiquitin E3 ligase for p53, while p53 is a transcription factor that binds DNA. While both proteins participate in the same biological process (regulation of transcription) and they physically interact with each other, there is no biochemical similarity between them (see Section 5.3 above).

Note, however, that two separate but largely redundant pathways should encode proteins that are largely overlapping in their molecular functions. This follows from the rationale that if two pathways are redundant, they must perform similar biochemical functions even if they have distinct participating proteins. This provides another functional test for synthetic lethality. Imagine a set of synthetic lethal pairs composed of two gene sets **A** and **B**, in which every member of **A** is synthetic lethal with every member of **B**, but there are no synthetic lethal pairs within **A** or **B**. If the Molecular Function GO positions of all genes in this set are enumerated, then one predicts that the average GO distance within **A** will be nearly equal to that within **B**.

A second prediction is that proteins encoded by **A** should preferentially physically interact among themselves, and so would proteins encoded by **B**, but that **A** proteins may not preferentially interact with those in **B** (see Figure 1.5a, where A and A' constitute examples of **A** and **B**).

There are many exceptions to the above tendency of mapping synthetic lethal pairs to between-pathway members (Kelley and Ideker, 2005), which may indicate incomplete knowledge of biochemical pathways or alternative interpretations of synthetic lethality. One alternative interpretation is that the products of a synthetic lethal gene pair could both belong to the same multi-protein complex (therefore belong to the same biochemical pathway; see Figure 1.5b). It may just so happen that the loss of one of them is insufficient to cause complete dysfunction of the complex, but that the simultaneous loss of both causes complete disruption. In this case, one expects to find an over-representation of protein-protein interaction edges among synthetic lethal pairs than expected by chance, and GO molecular function distances should be short among these synthetic lethal pairs.

5.8   Structure-Based Tests of Interaction

In Chapter 3, we discussed how structural properties of two proteins could be used to predict their propensity for physical interaction. Similarly, it is possible to test and score protein-protein interactions discovered by high-throughput methods if we know whether a pair of homologs of these proteins interact with each other and, if they do, the three-dimensional structure of the resultant complex. We have already seen how a protein-protein interaction network can be indirectly tested by comparison of the complexes that are *computationally predicted* from the network with experimentally discovered complexes. We now consider a more direct approach whereby two proteins that are found to interact in a high-throughput screen can be validated by computing a score that is based on the probability for *homologs* of these proteins to come in contact within a protein complex.

The basic method is summarized as follows. First, one constructs a database of experimentally discovered protein complexes that is sufficiently large to cover major functional families of proteins. From this database, one can identify pairs of amino acid residues (each member of the pair being from a different protein) that are in contact as defined by a suitable distance criterion. Amino acid pairs that are found to be frequently in contact among the protein complexes in the database are judged to have high propensity for interaction. It follows that a pair of proteins that has a large proportion of these amino acid pairs will be judged to have a higher propensity for interaction than pairs of proteins that have a low proportion of these amino acid pairs. Thus, a pair of putatively interacting proteins can be scored for their propensity for interaction according to the frequencies with which individual amino acid pairs within these two proteins are found to be in contact in the database of known complexes.

The methodology above was proposed by Aloy and Russell (2002). They developed a scoring function for measuring the propensity of interaction for an arbitrary pair of residues in terms of the ratio of the actual number (C_{ab}) to the expected number (E_{ab}) of contacts of those residues in the database. Specifically, for residues a and b, they defined the score S_{ab} by

$$S_{ab} = \log_{10}\left(\frac{C_{ab}}{E_{ab}}\right), \text{ with } E_{ab} = T\frac{n_a\, n_b}{N^2}, \tag{5.8}$$

where n_a and n_b are the total number of a and b amino acid residues in the database that participate in contacts, $N = \sum_i n_i$ is the total number of such residues, and T is the total number of interacting pairs of residues.

In Aloy and Russell's method, a putative interaction between two proteins is scored by finding the homologs of these proteins in the complex database and scoring the interaction between the homologs by summing up the scores of all interacting pairs of residues. This total (summed) score is then assessed for statistical significance by comparing this score to the total scores achieved by randomized versions of the protein sequences: if the actual score turns out to be larger than the scores of a large proportion of randomized sequences, the putative interaction is deemed significant.

This structural method of validating interactions is powerful: where homologs of proteins to members of known complexes exist, it not only validates but also accurately predicts the propensity for interaction. In addition, the method yields insight into the mechanism of interaction because it identifies the residues involved as well as the structure of the interaction surface. A drawback of the method, however, is its limited range of applicability at the time of writing: only a small fraction of the interactions discovered by large-scale experiments (yeast two-hybrid, co-IP, etc.) can be mapped to known complexes.

Chapter 6

Small Model Networks

In previous chapters we examined interaction networks of various kinds. The wiring diagrams, or topological properties, of known biological interaction networks form a small subset of all possible kinds of network organization. The most prevalent topological features are thought to be power-law degree distributions and modular structure, that is, the appearance of locally dense clusters, both of which we discuss in greater detail in subsequent chapters. These specific features of network topology might have been selected by evolution because they impart robustness to random attack, as discussed in Chapter 1. Alternatively, they might have no specific utility and be the result of a historical accident that is merely propagated by evolution. In either case, however, one would surmise that structural or topological features of the network should be informative of biological function. This is because the structure either had selective value or was preserved due to functional constraints on certain genes (or their products). In other words, if the topology itself had a selective property, then the study of the conservation of topological properties in evolution could potentially reveal the biological function that is selected for. On the other hand, if topological features merely happened to be preserved as a by-product of the conservation of some other features, then these other features might be revealed by studying the conservation of topology.

So far we have remained mostly silent on the biological relevance of network structure. Here we approach this issue by examining in detail a few specific models of biological interaction networks. These are probably the most well-understood network models in biology—the genetic and regulatory network controlling the life cycle of a bacterial virus (bacteriophage) called *lambda* and the cell cycle regulation of the lower eukaryotic organism (*Saccharomyces cerevisiae*). These discussions are aimed at illustrating the mechanistic basis of individual interactions and how these biological mechanisms are embedded within the network at large.

Box 6.1 Bacteriophages

A *bacteriophage* is a virus that can infect a host bacterium and reproduce within it. Bacteriophages are pathogens that are usually composed of a DNA or RNA genome that is packaged within a protein shell. Their highly reduced genomes encode a limited number of genes, approximately a third to a half of which are essential for the life cycle of the phage. There are several different types of bacteriophages, but here we focus on one special virus, the *lambda* phage.

6.1 Life Cycle of the Lambda Phage

The genome of bacteriophage (or simply, phage) *lambda* (Box 6.1) as it is contained in the free virus particles (virions) outside a virus-infected cell is a linear double-stranded (duplex) DNA with short single-strand ends that are complementary to each other. The phage particles adhere to a specific receptor protein (encoded by the host cell) that is expressed on the surface of a susceptible host *Escherichia coli* cell. Upon binding to the receptor protein, the DNA of the phage particle is injected into the host cell. The newly entered DNA quickly circularizes by complementary base pairing at its single-stranded ends, and a particular mode of DNA replication (the θ mode) begins, in which further circular DNA copies are made (Figure 6.1).

If only one phage infects the host cell, and this cell happens to contain abundant nutritional resources, it is observed that there is a $\sim$99% chance that the replication process will soon switch to a second mode (the σ mode; Figure 6.1), which leads to the production of numerous tandem copies of the virus genome. These tandem copies of phage genome together form the concatemeric linear double-stranded DNA[1], which is the substrate for the machinery that packages the DNA into the phage virion. Each cell produces approximately 100 to 300 virions, each of which contains exactly one genome's worth of lambda DNA. These virion particles emerge out of the host cell by destroying the host cell (lysis; see the bottom left panel of Figure 6.1).

The life cycle of the phage depends on the interactions between the

---

[1]A *concatamer* is a long stretch of duplex or double-stranded DNA that contains repeated copies of the unit genome.

phage-encoded proteins and the proteins of the host cell. For example, successful transition to the σ mode of replication requires the inhibition by phage-encoded proteins of certain host enzymes that normally degrade the linear double-stranded DNA. This inhibition is mediated by protein-protein interactions between the host enzymes and the phage-encoded proteins. Successful replication, expression of the virion proteins, and cell lysis are controlled by genes that are expressed by the phage genome late in the infection process. The entire process, beginning from the sigma mode of replication to emergence of the new virions, is the lytic cycle of lambda.

It is observed that if a host cell is simultaneously infected by two or three phages, the chance that the infecting phages will follow the lytic life cycle reduces to about 50% or 10%, respectively. The alternate mode of the life cycle—lysogeny—is the production of a set of proteins that allows the phage DNA to insert itself at a favored location on the host's chromosome (Figure 6.1). This is accompanied by repression of all genes encoded by the phage except one, the *cI* gene. The likelihood that the lysogenic cycle will be followed is somewhat increased if the host is growing under nutrient-poor conditions. In the lysogenic state, the phage DNA is called a *prophage* and the host cell that harbors the prophage is called the *lysogen*. As the lysogen divides, the prophage DNA replicates along with the host chromosome once every cell generation until there is a threat to the host genome (perhaps due to DNA damage, ultraviolet radiation, or heat shock). When this threat occurs, the phage DNA reverses the process of integration into the host genome and is released as a circular DNA of unit phage genome length. This circular DNA now undergoes the sigma mode of DNA replication and enters a lytic cycle. Subsequently the host cell bursts, thus liberating a crop of new virions capable of infecting other hosts. In the prophage state, the virus makes the host cell immune to further infection by related viruses, as we discuss below.

Remarkably, it is observed that the prophage state is extremely stable, such that a spontaneous conversion from lysogenic to lytic state has a probability of about 10^{-8} per phage particle per generation. Thus, if a single, recently established lysogen is followed at the single cell level, and given that a single host cell divides into two daughter cells approximately once every 30 minutes, a spontaneous switch from lysogenic to the lytic state would occur once every 5,000 years! This nonmutated state of the DNA in which all genes except *cI* are kept under the re-

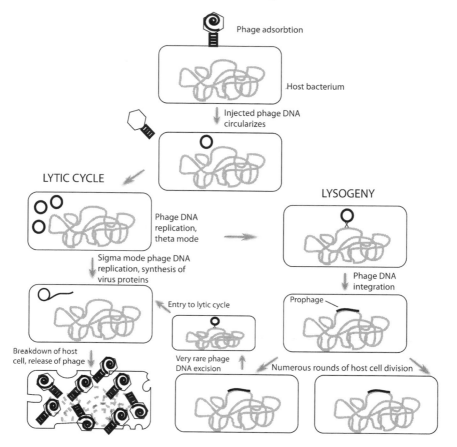

FIGURE 6.1: The life cycle of bacteriophage lambda and the lysis-lysogeny decision. Adapted with permission from Cooper (2000).

pressed condition is an example of the *epigenetic* state, which in this case appears to be more stable compared to spontaneous DNA base change due to mutation (Ptashne, 2005), a phenomenon that occurs at a rate of 1 in 10^6 to 10^8 per base-pair per generation. The mechanism that shuts down all phage genes except *cI* is also the basis of immunity against all further lambda virus infection, because the "early genes" of these newly invading genomes are also shut down by the cI protein.

We now examine how the remarkable biology of lambda is executed through the interaction of proteins, DNA, and RNA. One of the most important hallmarks of the life cycle of lambda is the decision to undergo either the lytic or the lysogenic mode, which has been classically described as a simple switch controlled by two proteins, cII and cIII.

Once this decision has been made, the state—either lysogenic or lytic—is maintained by the activity of one of two other proteins, cI or Cro, depending on whether the state is lysogenic or lytic, respectively.

6.2 Interaction Network in the Lysis–Lysogeny Decision

The essence of the interaction network of the type we discussed in previous chapters that is relevant to the lysis-lysogeny decision and maintenance is shown in Figure 6.2a. Each node in this network represents either a gene or an encoded protein, and edges between nodes represent protein-DNA or protein-RNA interactions. Gene names in this network are printed in italics while protein names are not. Early in the infection process, the host-encoded RNA polymerase protein activates the transcription of two phage-encoded genes, *N* and *Cro*. The product of gene *N*, namely the anti-termination protein N, binds to the RNA polymerase protein to form an RNA polymerase-N complex that activates transcription of *int, cII, cIII, O, P*, and *Q* genes. The O and P proteins are essential for starting the replication of lambda DNA, Int is essential for integration of the lambda DNA into the host chromosome once the decision to lysogenize has been made, and Q is another protein that modifies RNA polymerase by direct interaction. The RNA polymerase-Q complex allows transcriptional activation of the "lytic" genes (not shown in the figure). asQ is a negative regulator of *Q*, and cII is a transcriptional activator of *int, asQ*, and *cI*. The *cI* gene is both a negative regulator of *Cro* and *N*, and a positive regulator of itself. The cII protein is inactivated (degraded) by the host-encoded protein HflB. The cIII protein inactivates HflB. Thus we see that this network of interacting genes and proteins embodies several negative and positive feedback loops. In the next few sections, we study these feedback mechanisms further in the context of the life cycle of the lambda bacteriophage.

6.2.1 Regulatory Loops in the Lysis–Lysogeny Decision and Maintenance

The loops $N \rightarrow$ RNA Pol-N $\rightarrow cII \rightarrow cI \dashv N$, $cI \rightarrow cI$, and $cI \dashv Cro \dashv cI$ are required for the *establishment* of lysogeny. The path $N \rightarrow$ RNA Pol-N $\rightarrow$ CIII $\dashv$ HflB $\dashv$ CII $\rightarrow cI$ is also known to be important for establishing the lysogenic state (Figure 6.2b). Moreover, the positive

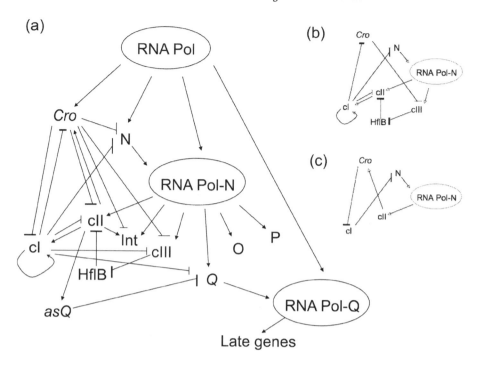

FIGURE 6.2: A highly simplified diagram of the gene regulatory network of phage lambda. (a) The networks important for three phases of the lambda life cycle (early, lysogenic, and lytic) are combined. (b) The subnetwork important for the establishment of lysogeny. (c) The subnetwork important for the maintenance of lysogeny. Edges with arrows denote activation (positive regulation) and edges with bars denote repression (negative regulation). Adapted and simplified with permission from Dodd et al. (2005).

feedback loop $cI \rightarrow cI$ is essential for the *maintenance* of lysogeny, and is the single mechanism responsible for the stable epigenetic prophage state discussed in the previous section.

The classical view of the lytic cycle induction is through the loop: $Cro \dashv cI \dashv N \rightarrow \text{RNA Pol-N} \rightarrow cII \rightarrow Cro$, leading to continued repression of cI by Cro (Figure 6.2c). The path $N \rightarrow Q$, initiated indirectly by Cro through the release of cI repression on N ($cI \dashv N$), is responsible for triggering the transcription of lytic genes. Note here that the relative dynamics of two effects of Cro on N, namely the magnitude and the rate of Cro's direct negative effect on N versus Cro's indirect positive effect on N via cI, is important for figuring out whether the expression of Q will be triggered, thus leading to a lysogenic state. This

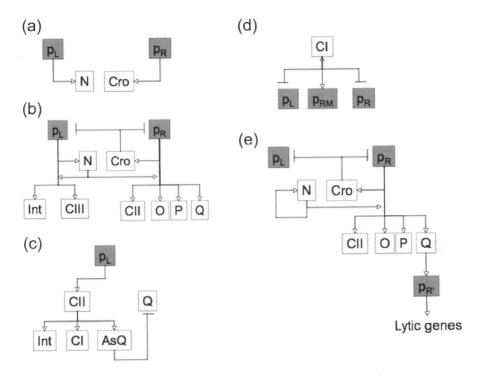

FIGURE 6.3: A highly simplified diagram of the temporal events in the gene regulatory network of phage lambda. (a) Early gene expression, immediately after infection; (b) delayed gene expression, before the lysis-lysogeny decision has been taken; (c) gene expression during the establishment of lysogen; (d) gene expression during the maintenance of lysogeny; (e) gene expression during the lytic phase. Note that stage (e) may follow immediately after stage (b) or (c). Shaded squares represent binding sites of the RNA polymerase and regulatory factors. For example, RNA polymerase binds to p_L to transcribe N, Int, and CIII, whereas CI binds to p_L to repress transcription from the position on the DNA. Adapted and simplified from Oppenheim et al. (2005).

cannot be captured by the static model represented by Figure 6.2; it must be augmented both by actual experimental data as well as by dynamical modeling of the expression levels of the various genes and proteins in this network. A more detailed summary of the timed expression of genes in the network is given in Figure 6.3, and a succinct description of these events is provided by Oppenheim et al. (2005), to which the interested reader is referred.

6.2.2   The Lysis–Lysogeny Switch

It is evident from the above discussion that competition between the two loops in Figure 6.2a, namely $N \rightarrow cII \rightarrow cI \dashv N$ and $Cro \dashv cI \dashv N \rightarrow cII \rightarrow Cro$, is the critical factor for tipping the balance between lysogeny and lysis. If a sufficient build-up of cII protein occurs, then the *cI* gene is activated, which turns off the expression of N and therefore *cII*, and consequently *Cro* is also turned off. Thus *cI* will reinforce its own synthesis through the $cI \rightarrow cI$ loop. If, on the other hand, *Cro* prevails, then *cI* is repressed while N remains active at least transiently (because it is found that the drop in cI concentration necessary to activate the expression of N precedes the rise in Cro concentration that represses N). This leads to $N \rightarrow Q$, which initiates transcription of the lytic genes. Note that the product of the Q gene modifies the RNA polymerase so that the latter can transcribe the lytic cycle genes without having to depend on N for its activity, as shown in Figure 6.2. These events lead to establishment of the lytic state.

The positive regulatory action of *cII* on its three target genes, *cI*, *int*, and *asQ*, requires an appropriate concentration of cII protein in the cell, which is determined by a number of factors, including the activity of the host-encoded protease HflB. Inhibition of HflB activity by cIII helps increase the accumulation of cII protein. Because it is observed that about 99% of the cells infected with one phage DNA undergo a lytic cycle, while only about 50% of cells infected with two phages do so, it is thought that more cIII and cII proteins are made in the latter case than in the former. The increased concentration of cIII and cII proteins is expected to be sufficient to trip the switch to establish a sustained amount of cI protein and hence to lysogenize. The HflB protein is present in relatively low amounts such that its inhibitory activity on cII can be overcome by cIII. Therefore it is likely that the path $cIII \dashv HflB \dashv cII \rightarrow cI$ should be critical for cI establishment. Two or more phage infections per cell are usually sufficient to provide enough cI for lysogeny.

One of the critical factors for the switch is the slight delay between the rise in cI expression and the activation of the Q protein. This is mediated by the positive regulatory effect of cII on transcription of *asQ*, the negative regulator of Q that we encountered earlier. This delay in the rise of Q protein expression, which is positively regulated by N (via RNA Pol-N) and also under feed-forward negative regulation through cII (via *asQ*), is currently thought to be critical in the establishment of lysogeny (Figure 6.2). This is because when N is transiently expressed due to a sudden drop in cI concentration, it triggers the expression of Q through the $N \rightarrow$ RNA Pol-N $\rightarrow Q$ path, while the repressive (negative feed-forward) path RNA Pol-N $\rightarrow cII \rightarrow asQ \dashv Q$ lags behind the direct induction of Q expression.

6.2.3 Stability of the Prophage

The cI protein is a highly efficient negative regulator (of *Cro* and *N*) as well as a positive regulator of itself (autoregulator). This remarkable property of a single protein arises due to a combination of its special structure and the geometrical arrangement of its binding sites on the DNA [see Court et al. (2007)]. Cro is transcribed from the p_L promoter while N is transcribed from the p_R promoter, both of which are repressed by cI (Figure 6.4). *Cro* expression is essential for late gene expression. Thus, a stable repression of *Cro* by cI maintains the stable prophage state.

What causes the unusual stability of the epigenetic state of repression of the lytic genes of lambda by cI? The cI protein has three important types of domains: one DNA binding domain, two dimerization interface domains, and one transcription activation domain. The DNA binding affinity of the monomeric protein is weak, but the simultaneous binding of two cI proteins on the two appropriately arranged binding sequences on the DNA allows protein-protein contact via the dimerization surfaces (Figure 6.4a,b). Cooperativity between the two monomers increases the stability of the bound dimer on the DNA target (Figure 6.4). End-to-end tetramerization of cI stabilizes the tetrameric cI-DNA complex further (Figure 6.4b). This prevents the promoters p_L (that controls *Cro* transcription) and p_R (that controls *N* transcription) from transcribing *Cro* and *N*.

The *N* and *Cro* genes are divergently transcribed, that is, they are encoded on opposite DNA strands, and have sufficient separation for the intervening DNA to fold. Each gene has three upstream tandem cI binding sequences arranged such that the tetrameric cI bound at two of the three binding sequences occludes the binding of RNA polymerase to the promoter sequences of the two genes. Furthermore, the tetramers bound at the two genes interact with each other, forming a highly stable octamer, with the intervening DNA looped between the two sites (Figure 6.4c). This looped structure, stabilized by the highly cooperative association of the octameric cI with its cognate binding

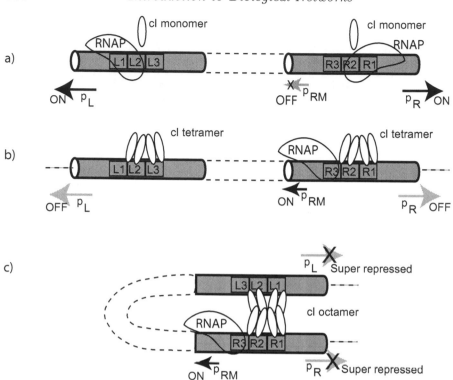

FIGURE 6.4: Cooperativity in protein recruitment on DNA. (a) The protein cI (ellipse) has weak affinity for each of the binding sites L1, L2, L3, R1, R2, and R3 on the DNA (represented by the cylinder). At low cI concentration, cI is unable to bind DNA, and RNA polymerase transcribes from the promoters p_L and p_R. However, p_{RM}, which transcribes cI, is inactive. (b) At higher concentrations of cI in the cell, protein-protein interactions reinforce each monomer's affinity for the binding site and allow two monomers to bind simultaneously to the DNA. This represses p_L and p_R but activates p_{RM} by allowing RNA polymerase to bind L3 through contact with the cI protein bound at R2. (c) Tetramer-tetramer interactions can fold the intervening DNA into a loop, and supercoiling of the DNA can provide additional stability to the structure. Adapted with permission from Ptashne (2005).

sequences, causes a practically complete repression of N and *Cro* gene expression.

A crucial element of the process of stable cI gene expression is the presence of a third cI target sequence near N and *Cro* (R3 and L3), somewhat further away from the two sites where the octameric cI binds. The cI protein has relatively less affinity for the third site than for the other two sites; when this site is unoccupied by cI, RNA polymerase is able to dock on the promoter region of the *cI* gene (p_{RM}) and allow transcription (Figure 6.4c). This allows continued production of cI protein for maintaining the repression of *N* and *Cro*, essential for a stable prophage state.

A relevant question here is how robust the lambda lysis-lysogeny circuit is to substantial changes in its component parts. It has been shown that despite drastic changes in the components (the replacement of R3 by R1, for example), little or no detectable change in the expression of any of the genes necessary for the switch occurred (Little et al., 1999). The apparent robustness of this circuit could not be predicted from past studies of the individual components alone.

6.2.4 Induction of the Lytic Cycle

The amount of cI protein required to maintain the lysogenic state is just sufficient to block the transcription of the *N* and *Cro* genes of any subsequent phage particles that may infect the lysogenic cell, thus explaining the immunity of the lysogenic cell to further infection. However, the maintenance concentration of cI can be reduced by factors such as DNA damage, in which case the cI protein undergoes degradation by an unrelated process. This leads to a transient drop in cI level beyond that which can retain the octameric complex at the two sites as explained in the previous subsection (Figure 6.4). As a result, *Cro* is transiently transcribed from the promoter p_L (see Figure 6.3 for schematic representations of the promoters that control the relevant genes). The Cro protein is a repressor that binds to the same sites as cI (but lacks cI's activating function), thus repressing *N*, *cI*, *cII*, and *cIII*. However, the feed-forward inactivation of Q by asQ is also inactivated (Figure 6.2a) because this feed-forward circuit is dependent on the positive regulation of *asQ* by cII. This causes transient expression of the Q protein, which triggers the default pathway that activates *Q* via the RNA Pol-Q complex (Figure 6.2a). The activation of Q trips the lytic pathway, leading to the transcription of genes necessary for the σ mode of DNA replication, synthesis of phage coat proteins, assembly of the phage particles, packaging of DNA into the particles, and destruction of the host cell.

6.3   Recruitment of Proteins as a Theme in Regulatory Networks

The overarching theme behind lambda gene regulation and the successful completion of the lambda life cycle is the selective recruitment of specific proteins at the correct location on the DNA. For example, the host RNA polymerase has strong affinity for the promoter sequences p_L and p_R, which are upstream of N and cII, respectively. However, RNA polymerase falls off the DNA shortly after transcribing N or before even transcribing cII. The N protein interacts with the RNA polymerase so that the modified RNA polymerase can transcribe past this point. Aided by the interaction of N with the RNA polymerase, transcription reaches cII and $cIII$. We have already seen how protein-protein interaction aids the recruitment of cI on the DNA to mediate repression as well as activation. Maintenance of the repressed state by a highly stable and compact looped structure of the DNA is possible due to recruitment of the repressor proteins. This theme of regulation by recruitment is general: the recruitment occurs due to modular interaction domains that can be removed and replaced artificially or naturally over the course of evolution to produce novel functions (modularity as a recurrent theme in biological networks is discussed in Chapter 8); all gene regulatory networks involve one or more of the mechanisms described above, although with significant variation in the details.

6.3.1   Regulatory Mechanisms Other than Passive Recruitment

In both prokaryotes and eukaryotes, transcription of most genes is activated by the recruitment of binding proteins that provide an "adhesive surface" on which the basal transcriptional apparatus can dock and begin its activity. There are, however, some exceptions to this in prokaryotes. In nitrogen metabolism genes, a form of the RNA polymerase spontaneously binds the promoter region but does not begin transcription until an enzymatic process activates the inactive transcription. This exception in prokaryotes is actually the rule in eukaryotes, where the RNA Polymerase II complex binds, often aided by passive recruitment by DNA-binding transcription factor proteins, to form the so-called "transcription pre-initiation complex" that must be "activated" first by a series of different enzymes. Only upon this "activation" step does the RNA Polymerase II molecule begin sustained synthesis of RNA upon the DNA template. Thus, passive recruitment provides the connectivity of the network and enzymatic modification triggers RNA synthesis.

6.4   The Network of Cell Cycle Regulation

We have seen how a relatively simple network of gene, protein, and RNA can regulate the complex life-cycle choices of a virus. We have also noted some of the difficulties in putting together models of these interaction networks that accurately describe the properties of the biological system. In this section we examine a somewhat more complex network of great practical value, the cell-cycle network of baker's yeast, and show that this network appears to have evolved a highly robust structure. The general lesson one learns from this is that biological networks are likely to have been selected for robustness and stability. If we are able to identify specific robust properties important for function, then it should be possible to select the most robust topologies as likely candidates for the actual network topology as well as explicitly design synthetic networks with a view toward endowing them with identifiable robust properties. In Chapter 8 we discuss properties of networks that appear to underlie robustness. Here, on the other hand, we discuss an important network that has been pieced together over the last several decades by biologists. The level of detailed understanding that we currently possess makes it now possible to model and simulate the behavior of this network for making quantitative predictions that are testable by genetic and biochemical experiments. Importantly, this network is highly conserved among all eukaryotic organisms, including humans, in which errors in the behavior of this network cause diseases such as cancer. Thus, quantitative modeling of the human cell-cycle regulation network, although not yet possible, will likely be crucial for understanding cancer and the effects of chemotherapeutic agents on cancer cells. Here, however, we restrict ourselves to a discussion of the cell-cycle network in yeast cells, which is simpler than its human counterpart.

Over 700 genes are involved in the control of a cyclic series of events that control the accurate progression of various landmark events in the life of a cell and the clock-like behavior underlying the normal growth and division of the single-celled eukaryotic organism, *Saccha-*

romyces cereviseae. This has become a model system to study processes of cell division in all eukaryotic organisms because the central mechanism and the genes that control these are highly conserved among eukaryotic organisms, as mentioned earlier. Using the yeast cell cycle as a model, crucial insights on cancer have been obtained. For example, most normal human cells with a well-defined function, such as an epithelial cell (which forms the lining of most tissues, including skin, blood vessels, ducts through which hormones are secreted, etc.) or a nerve cell or a muscle cell, do not divide. These specialized cells, also called differentiated cells, arise from so-called "progenitor cells" that have limited capacity for cell division and are themselves derived from "stem cells" that can readily divide. The undividing differentiated cells are demonstrably "arrested" at a specific cell-cycle phase, called G0. To re-enter the division phase, as occurs when a differentiated cell is transformed into a cancer cell, a series of genetic events must occur. Studies of the yeast cell cycle have contributed to our understanding of molecular events that must occur in these transitions. Another area where yeast cell cycle studies have contributed to understanding cancer is the mechanism by which cell-cycle progression is coupled to the completion of various essential cellular events during the life of cell. For example, the entire DNA of the genome must be replicated—a complex and somewhat slow process relative to the total time of the cell division cycle—before the nucleus divides into two daughter nuclei. Otherwise, drastic chromosomal abnormalities will occur, such as the ripping apart of incompletely replicated DNA duplexes during the nuclear division process. Cancers are generally associated with many such chromosomal abnormalities, thus suggesting that cancer cells are associated with defects in the biological processes that ensure the completion of DNA replication before nuclear division. Let us now see what these biological processes are.

Yeast cells are normally at a resting phase in which they grow in volume by metabolizing nutrients. This stage is called G1 (Figure 6.5). When a certain critical nuclear-to-cytoplasmic volume ratio is achieved, the so called START "switch" is actuated, so that the cell enters the S phase in which DNA replication is initiated and subsequently completed, which results in doubling the number of chromosomes. At that point the cell enters the G2 phase, which is a preparatory phase for the next phase, M or mitosis. In the M phase, the two sets of chromosomes undergo a complex series of associations, following which they

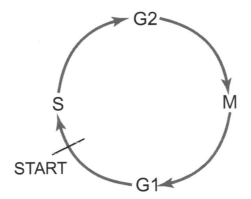

FIGURE 6.5: Sequence of major cell cycle events in baker's yeast. The cell cycle starts at the end of the G1 (resting) phase, when it enters the S phase. The S phase is characterized by replication of DNA and repair of discontinuities in DNA replication. This is followed by a second brief rest phase, G2. The cell then enters the M phase, when DNA becomes compacted into distinct chromosomes that assemble at the equator of the cell nucleus, tethered to the "spindle" apparatus. The homologous sets of chromosomes undergo separation, followed by movement of each complete set of chromosomes to one of the poles of the cell, and division of the cell cytoplasm into two daughter cells. At the end of M phase, the cells enter G1, during which metabolic activity leads to an increase in cell mass. When a critical cell mass (more accurately, a critical nuclear-to-cytoplasmic volume ratio) is reached, the cell leaves G1 to enter the S phase again. In the absence of a cell size signal, the cell continues to stay in the G1 phase.

move away to the opposite ends of cell, the cell divides in two, and each daughter cell receives one complete set of chromosomes. Then each cell reenters the G1 phase.

The cell cycle can be compared to the cycle of a washing machine, where a timing device controls the sequential progression of a series of separate events. If one of the processes is defective, the progression of the sequence of events is halted until the defective process is completed. Likewise, in the cell cycle, if DNA replication is incomplete, then the cycle arrests in S phase. If the chromosomes fail to align appropriately prior to separation of the pairs to the two daughter cells, the cycle arrests in M phase. If not enough nutrients are present or if another signal that is required to commence START is not initiated, the cell cycle arrests in G1 phase. These points where the cell cycle is arrested pending the completion of important cellular events are termed "checkpoints" in the cell cycle. Thus we speak of S- or M- or G1-phase-specific checkpoints. Sometimes one makes finer distinctions. Examples include DNA-damage checkpoints, signifying checkpoint arrest at the termination of S-phase or within the M-phase, in which the cell cycle is arrested due to damaged DNA left unrepaired; and the spindle checkpoint, which is actuated due to incomplete function of the spindle apparatus that pulls the chromosomal sets apart.

Unlike the washing machine that has a specific program to control the sequence of events, pumps, and sensors that monitor the completion of each event, the cell has no such physical program or clock, although it does have sensors. Instead, the cell relies on the dynamical properties of a regulatory network to control the precise timing of the cell cycle events.

A central role in this network is played by an enzyme called Cyclin-dependent kinase (CDK, also called Cdc28p in yeast). This enzyme has the catalytic property of adding high-energy phosphate groups to serine or tyrosine residues of many other substrate proteins, thus in turn affecting the chemical properties of the substrate proteins (many or most of these substrate proteins are enzymes themselves, whose catalytic properties are affected by the gain or loss of the high-energy phosphate group). The catalytic property of Cdc28p is altered by the binding to a set of cyclins, which are a set of small-molecular-weight proteins of related amino acid sequences. There are several different cyclins in every eukaryotic organism, including yeast. The various cyclins are synthesized in certain specific phases of the cell cycle, depending on the activities of phase-specific transcription factors that regulate the transcription of these various cyclin genes, and a set of negative regulators that degrade (break down) these specific cylin proteins in a cell cycle phase-specific manner. Thus, the G1 cyclin, for example, is synthesized in G1 phase due to accumulation of a specific transcription factor protein and the lack of a G1 cyclin degrading enzyme in the G1 phase. When the S phase begins, an enzyme that degrades G1 cyclin becomes active. Simultaneously, the transcription factor that triggers the synthesis of the G1 cyclin mRNA is no longer active. This combined process of the control of a cell phase-specific cyclin synthesis and degradation ensures a sharp peak of G1 cyclin activity during the G1 phase and its low activity in other phases of the cell cycle.

Out of approximately 700 different proteins, a core set of 11 proteins and their interactions capture the essential dynamics of the system, which is represented in the simplified model of Figure 6.6 (Li et al., 2004).

Many important processes in cell cycle regulation can be described as biochemical switches (with an ON state and an OFF state), a simplification that does not sacrifice the essential biology of the problem. Therefore, it is reasonable as a first approximation to assume that there are $2^{11}(= 2,048)$ possible states for this network of 11 nodes.

In a simple Boolean model of the network, the state S_i of the ith node at time $t + 1$ is a function of states at time t of all nodes that are connected to it, described by the following rule:

$$S_i(t + 1) = \begin{cases} 1, & \text{if } \sum_j J_{ij}S_j(t) + h > 0 \\ 0, & \text{otherwise,} \end{cases} \qquad (6.1)$$

where $J_{ij} = 1$ if node j is connected to i in an activating (or positively regulating) manner, $J_{ij} = -1$ if node j is connected to i in an inactivating (or negatively regulating) manner, $J_{ij} = 0$ if the two nodes are not connected, and h is an arbitrary threshold parameter. Assuming $h = 0$, that is, the genetic switch has no influence when it does not receive any input information, and is active when the input signals are present, the network can be simulated on the computer to start from a given initial condition (which is one among 2,048 possible initial conditions) and the states of each of its 11 nodes followed in discrete time steps to find stable configurations of the network as a whole, if there exist any.

In addition to the formulation of the rate laws of relevant protein concentrations in terms of a set of constants and parameters as described above, the model also uses conditional logic to reduce the set of possible interactions among the network components and for consistency with available biological experimental data. Due to the existence of the conditional logical steps and the large number of equations, the model is not solved analytically but is instead integrated numerically. In other words, the resulting differential equations and rules are integrated numerically to produce time course plots of protein concentrations that are considered important output functions of the model.

When this computer experiment was carried out with the simplified cell cycle model in Figure 6.6, 1,764 of the 2,048 possible initial states converged to a stable state of the network in which the values for the expression levels of Cdh1 and Sic1 were 1 (ON) each, and all the rest were 0 (OFF). This suggests that for most of the initial states, the

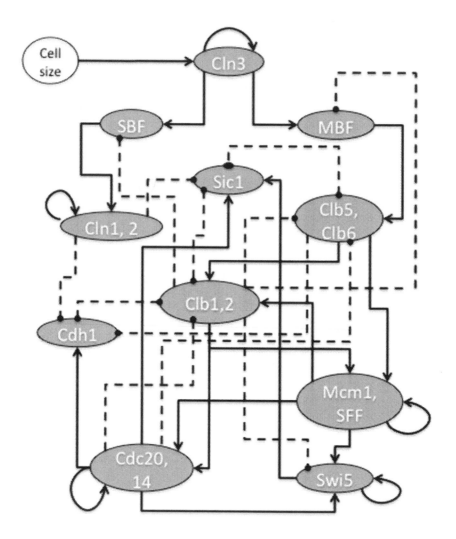

FIGURE 6.6: A simplified model of the network of cell cycle regulation. Each node represents either a single protein or a protein complex. Positive regulation is represented by solid lines ending in arrows and negative regulation by dashed lines ending in filled circles. Some edges are bidirectional. The node labeled "Cell size" denotes a collection of genes that produce a signal when the ratio of nuclear and cytoplasmic volumes drops below a certain threshold. This drop is characterized by a transition from G1 to S phase in the cell cycle; metabolic activities prior to the transition lead to an increase in cytoplasmic volume while the nuclear volume remains relatively constant. Adapted with permission from Li et al. (2004).

"flow" through the network converges to a network state in which, of the 11 proteins, only two (Cdh1 and Sic1) are expressed and the remaining proteins are not expressed. There were two other minor stable states, which could be reached from 151 and 109 initial network states, respectively. For all three stable network states, the final protein expression patterns corresponded to those in the G1 phase of the cell cycle. Furthermore, four additional stable states were found that could be reached from less than 10 initial states each. When random networks with 11 nodes and the same number of edges (in which the connections were randomized) were tested for stable states, no such bias for stable state configurations was observed.

The above observations show that the cell cycle regulation network, which undoubtedly has evolved through natural selection, has a single stable state corresponding to the G1 phase of the cell cycle. Noting that most environmental signals, such as low nutrition availability or a hormone signal, arrest cells in the G1 phase (cells in other phases complete their progress through the cell cycle, enter G1, and are arrested there), this simulation exercise provides strong evidence that the architecture of this network might have evolved to be robust to initial states. That the cycle maintains the biologically relevant progression can be tested by "exciting" the START state in G1 with the cell size signal, which then can be observed to progress through S, G2, M and return to G1 via discrete steps.

Given the small size of the core network of cell cycle regulation in Figure 6.6 as compared to the actual network, which despite its massive size [880 chemical species and 732 reactions (Kaizu et al., 2010)] still remains incomplete, it is surprising that the reduced model captures the essential properties of the cell cycle. The Boolean model of the cell cycle network described above, however, is insufficient to capture more subtle dynamics, such as the actual rates of increase and decrease of various protein levels and the timescale of their periodic variations over the course of the cell cycle. Ultimately, the aim is to develop a mathematical model of the cell cycle that can quantitatively predict the effects of variation in any of the protein levels on all other proteins and cell biological events. One use of such a model, for example, would be to predict the quantitative effects of a drug molecule that activates or inactivates an enzymatic step in the process, as in the case of an anticancer drug. To that end, it is necessary to construct more detailed models based on the mechanistic understanding of each of these steps.

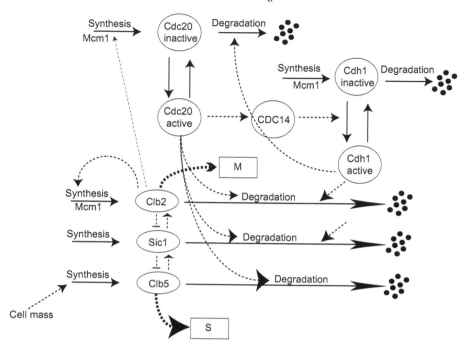

FIGURE 6.7: A partial mechanistic model of cell cycle regulation in budding yeast. Only a few reactions important for the control of DNA synthesis in the S phase and entry into the M phase are shown in boxes. These reactions, along with several more, were numerically modeled by Chen et al. (2004). Solid arrows denote chemical reactions while dashed arrows denote regulation.

In a particularly interesting example of quantitative modeling of the yeast cell cycle, a mechanistic model of the most essential enzymatic steps in cell cycle regulation was constructed (Chen et al., 2004) (Figure 6.7). This model had nearly twice as many nodes as that in the Boolean model discussed above. Corresponding to each enzymatic step in this model, a differential equation was written down, which resulted in a system of coupled differential equations. For example, corresponding to the rate of change of active Cdc20 protein concentration with time, which is positively regulated by Mcm1 and is also degraded at a constant rate through the action of Mad2, one writes

$$\frac{d[Cdc20]_T}{dt} = k_1 + k_2[Mcm1] - k_3[Cdc20]_T, \qquad (6.2)$$

where $[Cdc20]$ and $[Mcm1]$ represent concentrations of the proteins Cdc20 and Mcm1, respectively, and $k_{1,2,3}$ are rate constants for the

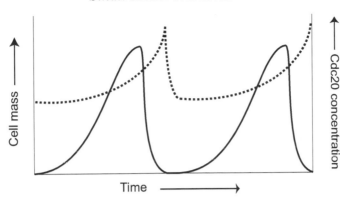

FIGURE 6.8: A schematic diagram of the results of simulation of a computational model of the cell cycle network of Figure 6.7. The dashed curve corresponds to the cell mass and the solid curve corresponds to the concentration of Cdc20.

synthesis and degradation of $[Cdc20]$. The values of the individual rate constants were either calculated by actual experiments or chosen to lie within reasonable ranges. The simulation output included cell mass and concentrations of certain sentinel proteins, such as cyclins or CDK activity. Simulation of this set of coupled differential equations yields time-dependent concentrations of the various sentinel proteins and cell mass (Figure 6.8), each of which shows a cyclic pattern. The sentinel proteins are those that exhibit periodic cell cycle-dependent variations of concentration in the cell and have been shown to cause the accurate progression of the cell cycle itself. Using the simulation data, it is possible to make precise predictions of the concentration dynamics given any perturbation to the model (representing experimental perturbation of a component of the cell cycle regulatory network). Alteration of parameters in the model can simulate deletions, gene dosage effects, binding affinity changes, etc. In Equation (6.2), for example, the value of the constant k_2 can be set to 0, signifying that the transcription factor Mcm1 fails to induce the transcription of the Cdc20 gene (in fact, the relevant protein is a complex of the Mcm1 protein and another transcription factor protein called SFF, which is omitted here for simplicity). We can then numerically simulate the effect of this adjustment on all other concentrations. Similarly, one can increase the value of k_2, which signifies an increase in the efficiency with which Cdc20 mRNA is synthesized, and then simulate its effects. Working in this fashion, it is possible to systematically search the parameter space for parameter values for which the species concentrations lie within normal bounds, thus pinning down the robustness regime of the yeast cell cycle network as modeled in this simplified version. Moreover, many of these predicted effects can actually be tested by genetic experiments (e.g., by deleting the binding site of Mcm1p at the promoter site of the *CDC20* gene, or by artificially hooking up the *CDC20* gene to a stronger promoter sequence, and so on). Thus it is possible to experimentally identify incorrect predictions and use this information to iteratively alter the network so as to ultimately bring the experimental observations into congruence with those predicted by the altered model.

Chapter 7

Tractable Models of Large Networks

What are the design principles of biological networks? Are there specific constraints that operate on their structure as they evolve? What are the mechanisms of network evolution and what properties of biological networks do these mechanisms preserve? Questions such as these are difficult to answer in general, but have led to a flurry of activity in the area of building model networks that mimic the large-scale connectivity properties of networks observed in nature. In order to compare model networks to real biological networks, we consider here various global statistics that describe network structure, including the degree distribution, clustering coefficients, and mean path lengths.

As we have discussed in earlier chapters, many biological interactions at the molecular level consist of physical binding events and/or chemical reaction events. We know that one way a gene regulatory interaction occurs is when a transcription factor protein binds to the promoter region of the gene that is being transcribed. A protein-protein interaction consists of physical binding of two proteins to form a complex. In a metabolic network, a directed interaction edge signifies the occurrence of a chemical reaction. In more abstract networks, on the other hand, edges may not have such a transparent interpretation but may still arise out of multiple binding/reaction mechanisms. For example, two genes that have a synthetic lethal interaction frequently lie on separate pathways of serial protein-protein interactions that are connected to each other at one end point.

The physico-chemical events that underlie a network are susceptible to changes due to gene mutations. Mutations alter the gene sequences (and protein structures) of the fundamental players (nodes) and therefore change binding energies and reaction rates. Mutations can also alter DNA sequences where proteins (such as transcription factors) may bind to regulate a gene. A series of mutations in such noncoding DNA regions accumulated during the process of biological evolution will therefore lead to rewiring of the network due to altered transcription factor binding. Old binding sites may be lost. New sites can emerge. Furthermore, genes can be duplicated over the course of evolution and the number of such duplicates can change as the organism evolves. Under this process, multiple genes may carry the same biological function and therefore the same sets of network connections. These multiple genes, in turn, are independently susceptible to evolutionary divergence by mutations, thus leading to further rewiring of the network. Such considerations lead to a set of putative rules for the evolutionary dynamics of a network, consisting of a mixture of node duplication, node specification, and rewiring processes. The precise mechanisms that determine how these processes are manifested at the network level, as well as the relative frequencies of occurrence of each process, are largely unknown, and different assumptions about the nature of these processes and their frequencies lead to strikingly different types of model networks.

In this chapter, we lay the foundations for modeling network growth and evolution by using well-studied theoretical models as examples. The aim here is to understand the origins of biological complexity at the network level, as well as to use the models to extract information about how evolution operates at the network level. To gain insight into these models, we begin with a discussion of regular, random, and small-world networks, the way they are "wired," and their properties. Because large networks are complex objects, it is important to extract a few statistical measures that capture their overall properties in order to facilitate comparison of real networks with model networks. The statistical properties that we focus on here are the mean path length through the network and the average clustering coefficient of the network.

We further discuss various classes of models of network evolution, including biophysical models that represent first steps in bringing large network models into the familiar domain of known molecular interaction mechanisms. We then turn to the question of evolutionary insights that we obtain when model networks and model network mechanisms are confronted with real data. Unless otherwise stated, we assume in this chapter that all networks under consideration are connected networks or connected components of larger, disconnected networks (Box 7.1).

Box 7.1 Network components.

A network is said to be *connected* if every pair of nodes can be connected by a path constructed from the edges that comprise the network. A network that is not connected (i.e., has at least one pair of nodes that cannot be connected by a path along the network) is a *disconnected* network. A disconnected network can be naturally split up into sub-networks that are, by themselves, connected networks but are not connected to each other. These sub-networks are called *components* of the original network. Any two nodes within a component can be connected by a path but nodes in different components cannot be connected to each other.

In the case of directed networks, an additional distinction is made. A directed network is *weakly connected* if the corresponding undirected network is connected. It is *strongly connected* if, for every pair of nodes u and v, there exists a directed path along the network from u to v and one from v to u. Consider, for example, a directed network with three nodes a, b, and c and two directed edges: an edge from a to b and an edge from c to b. This network is weakly connected because there is an undirected path between any two nodes. However, it is not strongly connected because there is no *directed* path along the network from a to c. Note that a strongly connected network is automatically weakly connected but the converse is not true.

7.1 Models of Regular Networks

We begin our mathematical exploration of network models by considering what is perhaps the simplest model network, a *regular network*, in which every node has the same degree r. A regular network containing n nodes, each of degree r, has

$$E = \frac{1}{2}nr \qquad (7.1)$$

edges, where the factor of $1/2$ arises because each edge is shared by two nodes. More generally, for an arbitrary network, $E = (1/2)\sum_{i=1}^{n} k_i$, where k_i is the degree of node i, and the sum runs over all n nodes in the network. The study of regular networks as one extreme, and their counterpart, random networks, as the other, is essential to the understanding of more realistic networks that have properties that are intermediate between the two extreme models.

Although regular networks appear to be simple, there are many distinct ways of wiring a regular network out of a given number of nodes

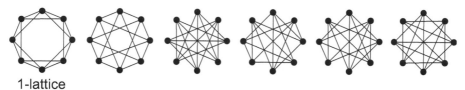

1-lattice

FIGURE 7.1: All distinct regular networks with $n = 8$ and $r = 4$, including the configuration that corresponds to a 1-lattice.

and edges. These distinct ways give rise to distinct networks. To be more precise, we introduce the concept of network *isomorphism*. We will say that two networks are isomorphic if there exists a one-one correspondence between the nodes of one network and the nodes of the other such that the number of edges (0 or 1) between any two nodes of one network is equal to the number of edges between the corresponding two nodes of the other network. Here, we tacitly assume that the nodes of the network have no labels, or even if they do, they are unlabeled for the purpose of establishing the isomorphism.

Thus, we say that two unlabeled networks are distinct if they are not isomorphic. The number of distinct labeled networks is smaller than the number of distinct unlabeled networks, because node labels essentially define the one-one correspondence between the two networks. The notion of isomorphism is quite general, and it is instructive to keep this in mind when comparing a network generated by a model to a real one. Figure 7.1 displays all distinct (i.e., nonisomorphic) regular networks with $n = 8$ and $r = 4$.

For our purposes, we consider a special type of regular network, called a 1-lattice, principally because the computation of many network statistics is relatively simple for such a network. A 1-lattice is a labeled, undirected network in which each node is connected to its immediately neighboring nodes in exactly the same way. Only one of the regular networks shown in Figure 7.1 is a 1-lattice.

Suppose we label the nodes on a 1-lattice with n nodes by the integers 1 to n. Mathematically, an arbitrary node in a 1-lattice, labeled v, is joined to all nodes u_i and w_i that satisfy the following criteria:

$$u_i = (v - i + n) \pmod{n},$$
$$w_i = (v + i) \pmod{n}, \tag{7.2}$$

where $1 \leq i \leq r/2$, and it is generally assumed that $r \geq 2$. The modulo operation above (abbreviated as "mod") ensures that the labels "wrap around" correctly. That is, the difference between the two sides of each equation in Equations (7.2) is an integer multiple of n. A 1-lattice can therefore be represented as a ring, as shown in the first panel of Figure 7.1.

7.1.1 Path Length in a Regular Network

It is quite straightforward to find the typical path length in a 1-lattice. Consider the mean length of all shortest paths between pairs of nodes in the network. Because all nodes in a 1-lattice are connected in the same way, let us start with some arbitrary node v. There are then exactly r nodes that have a shortest distance of 1 from v. For the sake of simplicity, we assume that r is even, so that there are equal numbers of nodes $(r/2)$ on either side of v. Because each of these r nodes also has degree r, there are a further r nodes that have a shortest distance of 2 from v. Proceeding inductively, there are an additional r nodes that have a shortest distance of 3 from v, and so on, until all the nodes in the network are exhausted. We can therefore group nodes in the network in such a way that two nodes belong to the same group if they have the same shortest distance from v. Each group contains r nodes, and if we do not assign the node v itself to any group, there are $(n-1)/r$ groups in the network[1]. We may label the groups according to their shortest distance from v. Thus, nodes in group i have a shortest distance of i from v, with i ranging from 1 to $(n-1)/r$. The average path length l is then equal to the sum of shortest distances from each group to v divided by the total number of groups:

$$l = \frac{\sum_{i=1}^{(n-1)/r} i}{(n-1)/r} = \frac{1}{2}\left(\frac{n-1}{r} + 1\right) \sim \frac{n}{2r} \text{ when } n \text{ is large.} \quad (7.3)$$

This shows that the average path length grows linearly with the size n of a 1-lattice. When the number of nodes is large, l is also expected to be large.

7.1.2 Clustering Coefficient of a Regular Network

The average clustering coefficient of a network is a measure of the density of triangles or cliques in a network. Alternatively, it is a measure

---

[1] We assume for simplicity that $(n-1)/r$ is an integer. If it is not, Equation (7.3) must be modified appropriately. This modification does not change the result that the mean path length grows linearly with the size of the network for large n.

of the probability that the direct neighbors of a given node are connected to each other. For each node, the *local clustering coefficient* of that node is the ratio of the actual number of edges between neighbors of the node and the maximum possible number of edges between the neighbors. Thus, if a node i has degree r, the maximum number of possible edges between its r neighbors is $E_{\max}(i) = r(r-1)/2$, and the clustering coefficient measures the actual number of edges as a fraction of $E_{\max}$.

An alternative definition of the local clustering coefficient can be given in terms of the number of triangles the node participates in, and is given by

$$C_i = \frac{N_\triangle(i)}{N_3(i)}, \tag{7.4}$$

where $N_\triangle(i)$ is the number of triangles that node i participates in, and $N_3(i)$ is the total number of connected triplets of nodes with i as the central node. That is, $N_3(i)$ is the same as the total number of pairs of edges emanating from i, which gives $N_3(i) = E_{\max}(i)$. Similarly, the total number of triangles that i participates in is just the number of ncighbors of i that are connected to each other. These observations show that the two definitions of local clustering coefficient that we have outlined are equivalent.

For the 1-lattice, the neighboring $r/2$ nodes on one side of an arbitrary node v are all connected to each other, resulting in $(r/4)(r/2-1)$ edges [because a set of m nodes in which every node is connected to every other node has $m(m-1)/2$ edges], and so are the neighboring $r/2$ nodes on the other side. Furthermore, there are an additional $(r/4)(r/2-1)$ connections between the neighbors on each side of v. This results in a total of

$$E = \frac{3r}{4}\left(\frac{r}{2} - 1\right) \tag{7.5}$$

connections among the r neighbors of v. The local clustering coefficient of v is therefore

$$C = \frac{E}{E_{\max}} = \frac{3}{4}\left(\frac{r-2}{r-1}\right) \sim \frac{3}{4}, \tag{7.6}$$

where the last, approximate relation holds for moderately large values of r.

Because every node in a regular network has the same value of C (from the definition of a regular network), the local clustering coefficient

TABLE 7.1: Typical Path Lengths and Average Clustering Coefficients for Various Networks

Network	Size	l	C
World Wide Web	153,127	3.1	0.11
Math. co-authorship	70,975	9.5	0.59
Silwood park food web	154	3.4	0.15
Word co-occurrence	460,902	2.7	0.44
E. coli reaction network	315	2.6	0.59
C. elegans neural network	282	2.7	0.28
Power grid (western U.S.)	4,941	18.7	0.08

Note: "Size" refers to number of nodes, and l and C are the typical path length and average clustering coefficient, respectively. *Source:* Albert and Barabási (2002).

is equal to the average clustering coefficient of the entire network. This average clustering coefficient does not depend on the size of the network and, as shown in Equation (7.6) above, is slightly less than 75% of its maximum possible value.

We have thus found two distinct topological properties of 1-lattices that are "global" in nature because they are computed using knowledge of the entire network. First, the typical path length scales in proportion to the size of the network and is therefore potentially large for large networks; and second, the average clustering coefficient is independent of the size of the network and is always fairly large. Of course, the terms "large" and "small" do not really mean anything unless we can compare these to corresponding values in other types of networks. Comparison of typical path lengths and clustering coefficients of real networks to those of the 1-lattice reveals that while average clustering coefficients of real networks are similar to those of a 1-lattice, the typical path lengths are far smaller (Table 7.1). We will now see that the opposite is true for random networks.

7.2 Random Network Models

Random networks, although they have a long history, have received much attention in recent years. They are useful both as models for man-made and natural networks as well as null models whose properties may be compared to those of real networks. In biology, random network models often form the basis for computing the statistical significance of an observation pertaining to an experimentally or computationally inferred biological network. On the other hand, the randomness in such models, it is hoped, can be made to capture the random component of the evolution of living systems. Thus, cleverly constructed random network models can potentially mimic the process of evolution at the network level, as shown through several examples in the following sections.

The seminal theoretical work of Erdös and Renyi (1959, 1960) [see also Bollobás (2001)] formed the foundation for the study of random networks. As such, a simple random network can be defined in two different ways that are essentially equivalent when the network is large.

The first way is to define a random network $G(n, M)$ as a labeled network with n nodes and M edges such that the edges are distributed uniformly at random between pairs of nodes. The second way is to define a random network $G(n, p)$ as a labeled network with n nodes in which each one of the potential $n(n - 1)/2$ edges exists with probability p. These two ways are practically equivalent if we make the correspondence $M = pn(n - 1)/2$. The distinguishing feature of a random network, in either case, is that the occurrence of an edge between two arbitrary nodes is independent of the occurrence of any other edge, that is, all edges are independent.

It turns out (as we shall see later in this section) that the simple concept of a random network as developed by Erdös and Renyi is sufficient to completely determine the distribution of node degrees (also called the *degree distribution*) of the network. The Erdös-Renyi random network is therefore not a good general model of the diverse types of network in existence, each of which may have different degree distributions, and is also not very useful as a null model that can be compared with a network that has a different degree distribution. For these reasons, a large amount of research has been directed toward extending the Erdös-Renyi notion of a random network to allow for arbitrary degree distributions and toward developing mathematical techniques for computing statistical properties of these general random networks. Before we embark on a more detailed discussion of the original Erdös-Renyi random network, it is thus instructive to learn the mathematical formalism for the description of general random networks and then treat the Erdös-Renyi random network as a special case of this formalism.

7.2.1 Random Networks with Specified Degree Distributions

We now consider random networks that have a prespecified degree distribution, which are constructed somewhat differently from the simple Erdös-Renyi random network. Our description follows that of Newman et al. (2001) and Molloy and Reed (1995, 1998).

In qualitative terms, a random network with a specified degree distribution will have node degrees whose frequency of occurrence mirror the probabilities specified in the degree distribution. These node degrees are therefore said to be *sampled* from the specified degree distribution. Each node degree is sampled independently of the degree of any other node. Because the degree of every node is sampled from the same degree distribution, we say that the set of node degrees constitutes an independent and identically distributed (iid) sample. We call this set of node degrees the *degree sequence* of the network.

Note that many different networks can have the same degree sequence; thus, the degree sequence alone does not uniquely specify the network itself. We therefore imagine an *ensemble* of networks that all have the same degree sequence. But the ensemble of networks consistent with the same degree distribution is even larger, because many different degree sequences can be sampled from a single degree distribution. Technically speaking, a *random network with specified degree distribution* is a network that is chosen uniformly at random from the larger ensemble.

When we compute network statistics, such as average path length and clustering coefficient, for random networks, we are really computing quantities that are averaged over the larger ensemble of networks described above. However, for reasons that are somewhat beyond the scope of this book, it turns out that if the network or the degree sequence is large enough, these statistics will accurately describe the statistics of a typical random network that has the specified degree distribution.

When forming a picture of a network with specified degree distribution, it is tempting to include a regular network in this picture because a regular network does have a specific degree distribution: one that is peaked at a single value! It is, however, important to emphasize that, unlike in the case of random networks, network statistics computed for a 1-lattice do *not* arise as a special case of the theoretical formalism presented below. This is because the 1-lattice displays a very special kind of wiring pattern that is not representative or typical of the wiring pattern of a randomly selected regular network. Thus, the theory described below is not really applicable to 1-lattices.

There are other networks that are not covered by the theoretical methods we are about to describe. These include networks in which the degree of a given node is not independent of the degrees of other nodes, as such networks cannot be generated by an iid sampling of a degree distribution. Many real networks have this feature of correlation between node degrees. However, the random network still serves a useful purpose here that we have indicated before—namely, as a statistical null model that can be compared to real networks.

7.2.2   Clustering Coefficient of a General Random Network

Consider a random network with some degree distribution p_k. That is, p_k is the probability that a node chosen uniformly at random from the network has degree k. We may define the *probability generating function* (pgf) $G_0(x)$ for the distribution p_k as follows:

$$G_0(x) = \sum_{k=0}^{\infty} p_k \, x^k. \tag{7.7}$$

Assuming that p_k is correctly normalized, it follows that $G_0(1) = 1$.

Given the pgf $G_0(x)$, we can obtain the original distribution p_k from it by taking appropriate derivatives:

$$p_k = \frac{1}{k!} \frac{d^k G_0}{dx^k} \bigg|_{x=0}. \tag{7.8}$$

The above equation explains why G_0 is called a generating function: derivatives of G_0 with respect to powers of x "generate" the degree distribution.

We may also use the pgf to compute moments of the degree distribution. Moments are mean values of powers of the degree. For example, the first moment, which is the average degree of a node, is given by

$$z = \langle k \rangle = \sum_k k p_k = G_0'(1), \tag{7.9}$$

where $\langle \, \rangle$ denotes the mean value, and a prime denotes a single derivative with respect to the argument. More generally, higher moments are given by

$$\langle k^n \rangle = \sum_k k^n p_k = \left[\left(x \frac{d}{dx} \right)^n G_0(x) \right]_{x=1}. \tag{7.10}$$

Another important property of the pgf arises from the following observation. If the distribution of some property l is generated by some generating function, then the distribution of the sum of different values of l, summed over, say, m independent realizations, is generated by the mth power of the pgf. For example, if we choose m nodes at random from a random network, then the distribution of the sum of their degrees is generated by $G_0(x)^m$. To see how this works, consider the expansion of $G_0(x)^m$:

$$G_0(x)^m = \left(\sum_k p_k x^k \right)^m$$

$$= \sum_{k_1 k_2 \ldots k_m} p_{k_1} p_{k_2} \cdots p_{k_m} x^{k_1 + k_2 + \cdots + k_m}. \tag{7.11}$$

The coefficient of x^k in the sum above is a sum of products of the form $p_{k_1} p_{k_2} \cdots p_{k_m}$ with $k_1 + k_2 + \cdots + k_m = k$, which is precisely the distribution of the sum of degrees of the m chosen vertices.

An important distribution that leads to the computation of mean path lengths and clustering coefficients for a random network is the distribution of the degree of nodes that we arrive at by following a randomly chosen edge. Such an edge will arrive at a particular node with probability proportional to the degree of that node. This node will therefore have a degree distribution proportional to $k p_k$. If we now consider starting at a randomly chosen node v and follow a randomly chosen edge starting at v and going to a nearest neighbor w, then the number of outgoing edges from w is distributed according to

$$q_k = \frac{(k+1)p_{k+1}}{\sum_j j p_j} = \frac{1}{z} (k+1) p_{k+1}, \tag{7.12}$$

where the $(k + 1)$ factor arises because w has degree $k + 1$ if it has k outgoing edges. The pgf for q_k is therefore

$$G_1(x) = \sum_{k=0}^{\infty} q_k x^k = \frac{1}{z} G_0'(x). \tag{7.13}$$

Consider a second neighbor u of v, and suppose that w and u have k_w and k_u outgoing edges, respectively. Under the independence assumption for random networks, the probability that u and w are connected is $k_u k_w/(nz)$, with k_u and k_w independently distributed according to q_k. The mean clustering coefficient of the network is then the mean of this probability, giving

$$C = \frac{1}{nz}\left[\sum_k kq_k\right]^2 = \frac{1}{nz^3}\left(\langle k^2 \rangle - \langle k \rangle\right)^2. \qquad (7.14)$$

Note that C is proportional to $1/n$, leading to the conclusion that large random networks with specified degree distribution have very small mean clustering coefficients. There are, however, exceptions to this when certain modifications are made to the definition of a random network. One of the modifications that is discussed in some detail below concerns random networks where not just the degree distribution but also the "cluster" (or triangle) distribution is specified.

Another example of a random network that displays large clustering is the so-called "one-mode projection" of a bipartite network. A bipartite network is one in which the nodes naturally split into two classes, say A and B, such that nodes of one class only connect to nodes of the other class. With a random bipartite network as a starting point, one can form a new network consisting exclusively of nodes in class A (or exclusively nodes in class B) such that two nodes in class A are connected if they shared a common neighbor in class B in the original bipartite network. This is a one-mode projection of the original bipartite network, and it has been shown to have high clustering coefficient (Newman et al., 2001).

7.2.3 Path Length in a General Random Network

Let us consider an arbitrary node u with degree k in a random network. As shown above, the generating function for the number of neighbors (other than u) of each of the k neighbors of u is $G_1(x)$. Therefore, using the "powers" property of the pgf outlined above, the pgf for the distribution of the total number of *second* neighbors of u is $G_1(x)^k$, assuming that the number of neighbors of each of the k neighbors of u are independent random variables[2]. Therefore, the pgf of the distribu-

---

[2]Because the neighbors of u can connect to each other, the independence assumption does not strictly hold. However, the assumption is a good approximation for large

tion of the number of second neighbors of a node picked at random is given by

$$\sum_k p_k G_1(x)^k = G_0(G_1(x)), \tag{7.15}$$

and, by similar arguments, the pgf of the distribution of the number of mth neighbors of a node picked at random is $G^{(m)}(x) = G_0(G_1(\cdots G_1(x)\cdots))$, where G_1 acts on itself $m - 1$ times. Thus the mean number z_m of mth neighbors of a randomly chosen node is

$$z_m = \frac{dG^{(m)}}{dx}\bigg|_{x=1} = G_1'(1)z_{m-1}. \tag{7.16}$$

With the initial condition $z_1 = z$, this recursion is solved as

$$z_m = \left(\frac{z_2}{z_1}\right)^{m-1} z_1. \tag{7.17}$$

We are now in a position to estimate the typical path length l in a random network. This path length is reached when the total number of neighbors out to that path length is equal to the total number of nodes in the network, that is, $1 + \sum_{m=1}^{l} z_m = n$. Using Equation (7.17), this gives

$$l = \frac{\ln\left[(n-1)(z_2 - z_1) + z_1^2\right] - \ln z_1^2}{\ln(z_2/z_1)}. \tag{7.18}$$

This shows that the typical path length grows logarithmically with the number of nodes in a large random network. Because of the logarithmic dependence of the path length on n, this path length can be quite small even for very large random networks. Table 7.1 shows that the typical path length in most real networks is quite small relative to the size of the network. This feature of real networks is therefore, at least in a qualitative sense, nicely captured by the random network model.

There are thus two important distinctions between the 1-lattice and a random network with an arbitrarily specified degree distribution. First, unlike a large 1-lattice, a large random network with specified degree distribution usually displays very little clustering (see, however, the caveat above regarding special cases where large clustering is possible in a random network as well as a generalization to the configuration model

---

random networks, in which the clustering coefficient (and hence, the probability that two neighbors of u connect to each other) is small.

discussed below). Second, typical path lengths in a random network are significantly smaller than in 1-lattices. These two distinctions will turn out to be important in the discussion of small world networks below. Before we do that, however, let us see how random networks with specified degree distribution can be constructed on the computer.

7.2.4   The Configuration Model and Its Modifications

Because random networks with specified degree distributions are useful as null models for testing hypotheses about the properties of real networks, it is important to be able to generate such networks on the computer in an efficient manner. One of the simplest methods of doing so is the configuration model of Bollobás (2001) and Newman et al. (2001). The method is quite simple and can generate either an undirected network with specified degree distribution or a directed one in which both ingoing and outgoing degree distributions are separately specified.

We first describe the more general case of generating a directed random network. In most practical applications, one is interested in comparing such a network to a real network in which both ingoing and outgoing degrees of every node are known. The construction of the randomized version of the real network then proceeds as follows. Suppose the real network has n nodes. For every node i in the real network with in-degree j_i and out-degree k_i, we create a node in the randomized network with j_i inward-pointing edge stubs and k_i outward-pointing edge stubs. Repeating this process for every node results in a collection of nodes with stubs. Each out-stub is then randomly paired with a distinct in-stub and the process is repeated until all stubs are paired. This procedure results in a random network that has the same set of in- and out-degrees as the real network under consideration[3]. It is common practice to further prune this network at the end by removing any multiple edges and self-loops, if the real network also does not possess multiple edges or self-loops. The case of an undirected network is treated in a similar manner, the only difference being that the stubs have no directionality and therefore any two randomly chosen distinct stubs can be paired.

Milo et al. (2002) describe a slightly modified version of the configu-

---

[3]This random network is, however, not uniformly sampled from the space of all networks consistent with the given degree sequence (King, 2004).

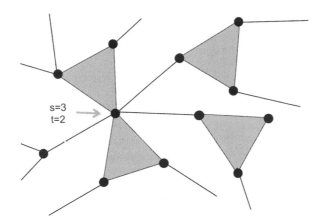

FIGURE 7.2: A piece of a network in which single edges and triangles are separately specified. The node indicated by the gray arrow has degree $k = 7$, with $s = 3$ single edges and $t = 2$ triangles. Note that $k = s + 2t$.

ration model presented above. In this version, a chosen pair of stubs is first inspected: if the addition of an edge across this pair would create a multi-edge or a sel-loop, a new stub pair is chosen; otherwise, the edge is accepted. If at some stage, no stub pair can be added without creating a multi-edge or self-loop, the entire sub-network constructed so far is rejected and the process is started anew from scratch.

It is also possible to construct random networks in which the local clustering around each node is specified in addition to the degree of the node, as shown by Newman (2009). The local clustering is specified by the number of triangles in which the node participates. Thus, in this generalized configuration model, one specifies two quantities for each node i: the number of triangles t_i in which the node participates and the number of single edges other than triangles, s_i. This setup is depicted in Figure 7.2. Construction of such a network on the computer proceeds in a manner analogous to that in the basic configuration model: to each node i we assign s_i single-stubs and t_i double-stubs. Then we randomly pick pairs of nodes and randomly connect single-stubs from each member of the pair. Similarly, we randomly pick triplets of nodes and connect randomly chosen double-stubs from the members of the triplet to form triangles. This is continued until no unconnected stubs are left behind. For this condition to be satisfied, the total number of single-stubs must be a multiple of 2, and the number of double-stubs must be a multiple of 3.

The generating function method may also be used to calculate network statistics for random networks with clustering. As an illustration of the method, we briefly outline the computation of the average clustering coefficient for such a network. In this computation, it is more convenient to use the triangle-based definition of the clustering coefficient. We have already seen that the local clustering coefficient of a node can be defined as the ratio of the number of triangles that the node participates in and the number of connected triplets that it participates in. From this definition, it can be shown that the average clustering coefficient of the entire network is (Newman et al., 2001)

$$C = \frac{3N_\triangle}{N_3}, \tag{7.19}$$

where $N_\triangle$ and N_3 are the total number of triangles and the total number of connected node triplets, respectively, in the entire network (the factor of 3 arises because a triangle corresponds to 3 connected node triplets).

Returning to the generating function method of computing the average clustering coefficient for random networks with clustering, let p_{st} be the joint distribution of singleton degrees and triangles for the network. A two-variable generating function $G(x, y)$ is then defined as

$$G(x, y) = \sum_{s,t=0}^{\infty} p_{st} x^s y^t, \tag{7.20}$$

in terms of which the ordinary generating function for the total degree may be expressed as

$$G_0(z) = \sum_{k=0}^{\infty} p_k z^k = \sum_{s,t=0}^{\infty} p_{st} z^{s+2t} = G(z, z^2). \tag{7.21}$$

We then have

$$3N_\triangle = n \sum_{s,t=0}^{\infty} t p_{st} = n \frac{\partial G(x, y)}{\partial y} \bigg|_{x=y=1}, \tag{7.22}$$

and

$$N_3 = n \sum_{k=0}^{\infty} \binom{k}{2} p_k = \frac{1}{2} n \frac{\partial^2 G_0(z)}{\partial z^2} \bigg|_{z=1}. \tag{7.23}$$

The derivatives on the right-hand sides of the two equations above cannot be computed explicitly without specific knowledge of the distribution p_{st}. However, it is important to note here that the factors of n cancel upon substitution into Equation (7.19). Thus, it is possible to have nonzero clustering even in the limit $n \to \infty$ for such networks, a feature that does not exist in typical random networks with specified degree sequence.

7.2.5 Erdös–Renyi Random Networks

As stated earlier, the Erdös-Renyi random network is completely specified by the number of nodes n and the parameter p that represents the probability that two nodes chosen at random are connected by an edge. This network corresponds to the standard notion of a random graph in the mathematical literature, and there are a large number of results for such graphs, most notably the dramatic emergence of a large connected component as p is increased [see, for example, Erdös and Renyi (1959, 1960); Bollobás (2001)].

Because there are $\binom{n}{k}$ ways in which a node can connect to k other nodes (out of a maximum of n connections), the degree distribution of an Erdös-Renyi random network is binomial:

$$p_k = \binom{n}{k} p^k (1 - p)^{n-k}, \tag{7.24}$$

giving

$$G_0(x) = e^{z(x-1)} \quad \text{and} \quad G_1(x) = G_0(x), \tag{7.25}$$

where $z = np$. Using Equation (7.14), this degree distribution gives a clustering coefficient of $C = z/n$, and, using Equation (7.18), a typical path length of $l \simeq \ln[n(z-1)]/\ln(z) - 1$ for large networks. Again, note the complementary properties of the 1-lattice (see Equations (7.3) and (7.6)) and the Erdös-Renyi random network with respect to clustering and typical path length: a random network has lower clustering and lower typical path length than a 1-lattice with the same number of nodes.

7.2.6 Small-World Networks

Many biological networks display some surprisingly similar global or topological features shared among many natural and man-made networks (Solé and Pastor-Satorras, 2003; Albert and Barabási, 2002). In Chapter 1, we briefly mentioned the prevalence of power-law degree distributions in protein interaction networks—a feature that is strikingly universal across many different natural and man-made networks—and we turn to some models that give rise to this feature in the sections that follow. It turns out that many biological networks that have power-law degree distributions, including protein interaction networks (Jeong et al., 2001), metabolic pathway maps (Jeong et al., 2000), and food webs (Montoya and Solé, 2003), also have *small* typical path length and *large* average clustering coefficients (Table 7.1). The simultaneous presence of these features is not consistent with either 1-lattice (or related) models or with standard random network models (unless, of course, one puts large clustering into a random network "by hand," as in the generalization to the configuration model we discussed above).

Networks that display small typical path lengths accompanied by large clustering, regardless of their degree distributions, are called *small-world networks*, following the seminal work of Watts and Strogatz (1998).

A simple method (Watts and Strogatz, 1998) of generating a small-world network exploits the fact that it is related to the 1-lattice with respect to its clustering and to an Erdös-Renyi random network with respect to its typical path length. One begins with a sparse 1-lattice ($r \ll n$) in which each edge is randomly rewired with probability p such that self-loops and multi-edges are excluded. This process introduces, on average, $pnr/2$ "shortcuts," which connect nodes that would otherwise be separated by a large path length in the regular 1-lattice. One can then examine the properties of the resulting network as a function of the single parameter p. The limit $p = 0$ corresponds to a 1-lattice, and the limit $p = 1$ corresponds to an Erdös-Renyi random network (Figure 7.3). While the Watts-Strogatz mechanism for generation of small-world networks is likely too simplistic to be an important biological evolutionary mechanism for small-world networks (at least in part because degree distributions of small-world networks created by this mechanism do not correspond to those observed in biological networks), it illustrates an important point: namely, that a very small amount of disorder (in the form of random rewiring) is sufficient to create "shortcuts" between distant nodes, and thus to create small-world networks out of regular ones.

Remarkably, one finds a range of values of p that exhibits small typical path length and large clustering, that is, small-world behavior (Figure 7.4). This simple model demonstrated for the first time that small path length could theoretically coexist with large clustering, in agree-

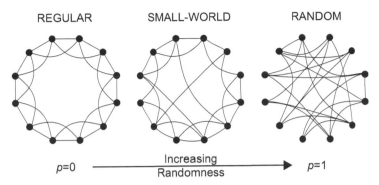

FIGURE 7.3: A small-world network (middle panel) can be generated by choosing a value of the rewiring probability p between the regular network limit ($p = 0$; left panel) and the random network limit ($p - 1$; right panel).

ment with the observed characteristics of real networks. The model is also consistent with (and indeed inspired by) the structure of some social networks, in which it is common to establish connections with many people in one's immediate vicinity, plus a few connections (short-cuts) with acquaintances far away that link separated neighborhoods.

7.3 Evolving Networks by Preferential Attachment

In addition to possessing small-world properties, many networks are found to have power-law degree distributions, that is, p_k is proportional to $k^{-\gamma}$, where γ is called the power-law exponent, as we briefly mentioned in Chapter 1. These networks include a variety of social networks, the World Wide Web, Internet domain and router networks, and disparate biological networks, including the network of protein-protein interactions in various organisms, metabolic networks, transcriptional regulatory networks, and protein domain networks [see, for example, Albert and Barabási (2002) for properties of these networks]. Such power-law distributions are not explained by the regular, random, or small-world network models we have studied so far[4].

---

[4]A degree sequence that is distributed according to a power-law distribution can be specified in constructing a random network according to the configuration model, but such a distribution is an input to the model rather than being a consequence of

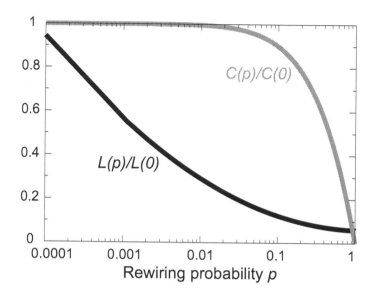

FIGURE 7.4: A schematic diagram of the typical path length (black curve) and average clustering coefficient (gray curve) for a Watts-Strogatz small-world network as a function of the small-world rewiring probability p (Watts and Strogatz, 1998). The two statistics are normalized by the values they take in a regular network: $L(0)$ and $C(0)$. There is a large range of values of p for which the typical path length is small and the average clustering coefficient large.

While networks that possess power-law degree distributions are termed "scale-free" networks, the latter phrase leads to many erroneous interpretations, including the implication that the network is, in some sense, self-similar, or that it resides at the critical point of a phase transition[5]. To avoid such interpretations, we refer to such networks as "power-law networks."

Although power-law distributions are ubiquitous in various contexts and have a long history, Barabási and Albert (1999) were the first to construct a simply interpretable dynamical model of network growth that robustly leads to a power-law degree distribution in a way that

it.

[5]For a fuller description of the history and possible mechanisms for power-law networks, the reader is referred to the essay by Keller (2005).

does not depend on the structure of the initial network. This model has been termed the BA model and is arguably the simplest dynamical model of network growth by preferential attachment of nodes. As we argue below, this model is also too simplistic to be a biologically plausible model for network evolution by itself, but it is such an important progenitor of biologically plausible network models of increasing sophistication that it is necessary to first understand the BA model in order to follow developments in the field.

The input to the BA model is a small network that can be quite arbitrary. The BA model is a dynamical model that evolves this initial small network to a larger network. For simplicity, "time" in the BA model (and in a host of other models that followed it) is imagined to proceed in discrete steps. A single new node with exactly m stubs (where m is chosen arbitrarily but remains fixed during the entire dynamics) is introduced into the network at each time step. Each of the m stubs preferentially link to existing nodes in the network, with probability proportional to the degree of the existing node. Thus, existing nodes with high degrees have a greater chance of being linked to newly born nodes; and conversely, a newly born node has greater chance of being linked to nodes that already have a large number of connections. This "rich getting richer" effect is characteristic of many social systems and is also plausible for networks such as the World Wide Web, in which new webpages usually preferentially contain links to existing pages that have high visibility. Note that in the BA model, no nodes or edges are deleted (thus, the network grows as time progresses), and the number of nodes in the network at time step t is equal to $t + m_0$, where m_0 is the number of nodes in the initial network.

With this picture in mind, we are now in a position to construct a mathematical description of the BA model. Suppose there are $n_{k,t}$ nodes of degree k at time t. After one time step, a new node of degree m has entered the network, thus increasing the number of nodes of degree m by 1. Further, m existing nodes have been chosen to be linked to the newly born node; the degrees of these m nodes have increased by 1. The probability of node i being chosen is proportional to its degree k_i. Because some node must be chosen, the sum over all nodes of this probability must be 1. Thus the properly normalized probability of node i being chosen is $k_i / \sum_j k_j = k_i / \sum_k (k n_{k,t})$. Putting all this together, the number of nodes of degree k at time $t + 1$ is given by

$$n_{k,t+1} = n_{k,t} + \delta_{k,m} + \frac{m}{\sum_{k'} k' n_{k',t}} \left[(k-1)n_{k-1,t} - k n_{k,t} \right], \qquad (7.26)$$

where the $\delta_{k,m}$ term represents the increase by 1 of nodes with degree m, and the remaining terms represent the increase in the number of nodes with degree k due to the increase in degree of nodes that earlier had degree $k - 1$, and the decrease in the number of nodes with degree k due to the increase in degree of nodes that earlier had degree k.

We now develop what is called a *master equation* for the probability $p_{k,t}$ of finding a node with degree k at time t, given by $p_{k,t} = n_{k,t}/(t + m_0)$. Rewriting Equation (7.26) in terms of this probability leads to

$$p_{k,t+1} = \frac{1}{t + m_0 + 1} \left[(t + m_0)p_{k,t} + \delta_{k,m} + \frac{m}{\langle k \rangle_t} \left((k-1)p_{k-1,t} - k p_{k,t} \right) \right],$$
$$(7.27)$$

where $\langle k \rangle_t \equiv \sum_{k'} k' p_{k',t}$ is the mean degree of the network at time t. Upon following the time progression of the above equation, it can be shown that the distribution $p_{k,t}$ becomes stationary on long timescales, when the network has become very large. At this point, $p_{k,t}$ can be assumed to be the constant, time-independent distribution p_k. It follows that the mean degree of the network also becomes constant at late times. Substituting p_k in place of $p_{k,t+1}$ and $p_{k,t}$, and substituting $\langle k \rangle$ in place of $\langle k \rangle_t$ in the above equation, we obtain, after some rearrangement of terms, the following relationship between p_k and p_{k-1}:

$$(k + \alpha)p_k = (k-1)p_{k-1} + \alpha \delta_{k,m}, \qquad (7.28)$$

where $\alpha = \langle k \rangle / m$. Multiplying the above equation by k on both sides and subsequently summing over all values of k leads to a simple expression for the mean degree at late times, namely, $\langle k \rangle = 2m$, which implies $\alpha = 2$. Substituting this value of α back into Equation (7.28) and rearranging terms again leads to a recursion equation for p_k in terms of p_{k-1}:

$$p_k = \frac{(k-1)}{(k+2)} p_{k-1}, \quad \text{for } k > m. \qquad (7.29)$$

The above recursion equation can be solved by using it to express p_{k-1} in terms of p_{k-2}, then p_{k-2} in terms of p_{k-3}, and so on, until we get to p_m. Thus, we obtain

$$p_k = \frac{(k-1)(k-2)\cdots m}{(k+2)(k+1)\cdots(m+3)} p_m = \frac{(m+2)(m+1)m}{(k+2)(k+1)k} p_m. \qquad (7.30)$$

For large values of k, the degree distribution p_k of the BA model is therefore proportional to k^{-3}, corresponding to a power-law distribution with exponent $\gamma = 3$. Note that this result does not depend on m or on properties of the initial, small network.

The BA model of network growth by preferential attachment gives a simple mechanism for generating networks with power-law degree distributions. Although the original form of the BA model outlined above can only explain power-law degree distributions that have an exponent of 3, there are generalizations of the BA model, based on the same fundamental idea of preferential attachment, that lead to a range of values for the exponent. A very simple model (Chung and Lu, 2006) begins with two preferential rules of growth: (1) node growth, in which a new node is added and connects with an existing node with probability proportional to the degree of the existing node, and (2) edge growth, in which a new edge is added that preferentially links nodes of high degree. At each time step, rule (1) is selected with probability p and rule (2) with probability $1 - p$. This model leads to a power-law degree distribution with exponent $\gamma = 2 + p/(2 - p)$. As p ranges from 0 to 1, the exponent takes values between 2 and 3. Similarly, a preferential attachment model (Dorogovtsev et al., 2000) that is slightly generalized from the BA form to include an initial "attractiveness" of existing nodes leads to exponents that can fall anywhere in the range $[2, \infty)$, depending on the initial attractiveness.

The preferential attachment mechanism appears a very plausible one for such networks as the World Wide Web, citation networks, and other types of social and man-made networks. In these networks, new nodes are likely to connect to existing nodes that already have a large number of connections, either to increase attention to themselves (World Wide Web and social networks), or because highly connected nodes have wide-ranging impact (citation networks). It is, however, not clear that the preferential attachment scheme is directly relevant to molecular networks in biology. There is no sound biological basis for why new genes or proteins that appear over the course of evolution must interact with other genes or proteins that already have a large number of interactions. Using yeast protein interaction data culled from two-hybrid studies, Wagner (2003) found a strong positive correlation between the degree of a protein and its likelihood of acquiring new interactions, a finding that is consistent with preferential attachment, albeit using high-throughput data that, as we know from earlier chapters, are error-prone. On the other hand, Middendorf et al. (2005) developed machine learning methods to find the best evolutionary model for a given network among a set of models. When applied to the *Drosophila* protein interaction network, their method rejects the preferential attachment model in favor of a duplication-mutation-complementation model that we discuss later in this chapter.

In spite of the possible limitations of the preferential attachment scheme when applied to networks in molecular biology, the reader should be aware that the BA model spawned a large amount of work on *evolutionary* models for networks, including many that did not use preferential attachment as a fundamental mechanism of growth. Specifically, more realistic models for networks in molecular biology are often based on the premise of gene duplication, a phenomenon that is known to have wide-ranging impact on the evolution of genomes.

7.4   Evolving Networks Based on Gene Duplication

Gene duplication is one of the primary forces behind the evolution of genomes. At the molecular level, gene duplication is a random event; it may occur as an error in DNA replication, during homologous recombination, by replicative amplification of transposable elements, or even arise by duplication of an entire chromosome. Usually, the duplicated version of a gene is under less selective pressure than its parent, and is therefore free to mutate rapidly and could potentially take on a novel function. This phenomenon results in *asymmetric divergence* over the course of evolution of the two duplicate members. Duplicated genes are called *paralogs*. Although they started as identical copies of the same gene, most paralogous pairs code for proteins that differ in structure and function because of mutations that have occurred in at least one member of the pair.

Duplication events have contributed to the emergence of novel functions in many species [see the review by Taylor and Raes (2004)]. For example, more than 33% of the *Mycobacterium tuberculosis* genome is comprised of recently duplicated genes. In a landmark analysis of 106 bacterial genomes, Gevers et al. (2004) found that between 7% and 41% of the genes had paralogs within the same genome, and that the genome size of a species or strain was strongly correlated with the number of paralogous genes it contained. There is strong evidence of ancient whole-genome duplication of the budding yeast *Saccharomyces cerevisiae* (Wolfe and Shields, 1997; Kellis et al., 2004), the plant *Arabidopsis thaliana* (Blanc et al., 2000; Vision et al., 2000), fish (Taylor et al., 2003), and two large-scale duplication events in the vertebrate lineage (Larhammer et al., 2002; McLysaght et al., 2002). In yeast, it is estimated that there are 10^{-3} duplication events per gene per million years (Lynch and Conery, 2000) and more generally in eukaryotes it is estimated that there is about one *whole-genome duplication* event every 100 to 200 million years (Evlampiev and Isambert, 2007). A study of the human genome sequence by Bailey et al. (2002) concluded that about 130.5 million base-pairs of the human genome (out of a total of just over 3 billion-base pairs) have been recently duplicated.

How does gene duplication manifest itself at the level of network evolution? It seems reasonable to assume that duplication of a gene can be modeled by duplication of the corresponding node on a network, where node duplication entails making an identical copy of the original node and all of its connections. Thus, we first examine the type of network growth that results from the action of such a duplication process by itself, and then consider the implications of duplication (of individual genes as well as sets of genes) acting in concert with other processes that model sequence mutations at the network level.

7.4.1   Network Growth via Node Duplication Alone

Consider a network that evolves only by node duplication, without any evolutionary divergence between duplicated nodes that may be arise due to mutations. While such an evolutionary model is unrealistic, we find that there is an important lesson to be learned from this idealized model, namely that because of the absence of random mutations, the network retains evolutionary memory of its original state (Raval, 2003).

We assume that the network under consideration is an undirected one and has m_0 nodes at some arbitrary initial time $t = 0$. At each time step, an existing node is picked at random and duplicated, that is, a new node is added to the network and the new node is connected to exactly the same set of nodes as an existing, duplicated node. Like the BA model, the number of nodes increases by one at each time step so that the total number of nodes in the network at time t is $t + m_0$. Consequently, the theoretical maximum degree in the network at time t is $t + m_0 - 1$ (where we assume that nodes do not self-connect). Because every node has equal probability of being duplicated at a given time step, the probability $p_{k,t}^{\text{new}}$ that a newly created node has degree k at time t is given by

$$p_{k,t}^{\text{new}} = p_{k,t-1}, \tag{7.31}$$

where $p_{k,t-1}$ is the degree distribution of the network at time step $t-1$. Furthermore, the probability $p_{k,t}^{\text{ndup}}$ that a node of degree k is a neighbor of a duplicating node is proportional to its degree, and a node that has the maximum possible degree will be a neighbor of a duplicating node with probability 1. This gives

$$p_{k,t}^{\text{ndup}} = \frac{k}{m_0 + t - 1}. \tag{7.32}$$

The two equations above enable us to determine the average number of nodes $n_{k,t}$ that have degree k at time t in a recursive fashion:

$$n_{k,t} = p_{k,t}^{\text{new}}\left(n_{k,t-1} - 1\right) + \left(1 - p_{k,t}^{\text{new}}\right)n_{k,t-1} + p_{k-1,t}^{\text{ndup}}n_{k-1,t-1} - p_{k,t}^{\text{ndup}}n_{k,t-1}, \tag{7.33}$$

where the first two terms represent the contribution to the number of nodes of degree k by the newly created duplicate node, and the next two terms represent this contribution due to the change in degree of neighbors of the duplicated node. Noting that $p_{k,t} = n_{k,t}/(t + m_0)$, the above equation may be converted to a simple master equation for $p_{k,t}$:

$$p_{k,t} - p_{k,t-1} = \frac{k-1}{t + m_0}p_{k-1,t-1} - \frac{k}{t + m_0}p_{k,t-1}. \tag{7.34}$$

The above equation actually holds for all nonisolated nodes (nodes with degree > 0). If there are isolated nodes in the initial network, their duplicates will also be isolated. Indeed, the proportion of isolated nodes in the network remains constant under the pure duplication dynamics envisioned here. Therefore, it is reasonable to consider solving Equation (7.34) with the assumption that the network contains no isolated nodes. In fact, Equation (7.34) represents one of the few evolving network degree distributions that is analytically solvable for all t (Raval, 2003). The large-time limiting behavior of this solution is, for $m_0 \ll k \ll t$,

$$p_{k,t} \sim \frac{m_0}{t}\binom{m_0 - 1}{k_{\min} - 1}\left(\frac{k}{t}\right)^{k_{\min}-1}p_{k_{\min},0}, \tag{7.35}$$

where $k_{\min}$ is the smallest (non-zero) degree value in the initial network. Thus the degree distribution at late times exhibits power-law behavior, albeit one in which there are *more* nodes with higher degree than nodes with lower degree. Perhaps more importantly, the degree distribution, unlike in the BA model of preferential attachment, depends on the nature of the initial network.

This feature of "initial condition dependence" of the degree distribution appears to be characteristic of network evolutionary models in which node duplication is the predominant mechanism of network growth. It is tantalizing to imagine the possibility of *unambiguously* inferring aspects of the topology of gene networks as they were millions of years ago from present-day data. However, in order for such a possibility to be realized, it is necessary to first have a systematic procedure for iteratively constructing and validating large-scale network models. It is also interesting to note that, for all genomes that have been sequenced so far, the fraction of genes having paralogs has been consistently less than 50%, a finding that would appear to refute the possibility of gene duplication being the dominant mechanism of genome and, consequently, network evolution. In reality, though, there is no simple and consistent method to infer the probability of node duplication at the network level from the extent of paralogy, although the topic is being actively researched. We discuss some aspects of this issue later in this chapter. In general, reconstruction of ancient networks is an important field of study but in the absence of duplication-dominance, it is only possible with a number of simplifying assumptions (see Section 7.5).

We now consider an important "mixed" model in which node duplication is not the sole dynamical mechanism of network growth, even though it forms an important component of the dynamics.

7.4.2   A Duplication–Mutation–Complementation Model of Network Growth

A model proposed by Vázquez et al. (2003) for the evolution of protein interaction networks is inspired by the idea (Force et al., 1999; Lynch and Force, 2000) that, subsequent to a duplication event, both copies of a duplicated gene are subject to degenerative mutations and lose some functions, but jointly they retain the full set of functions present in the ancestral gene. This observation can be translated into an evolving network model that grows according to the following rules at each time step, as shown in Figure 7.5:

(a) *Duplication step*: A node i is selected at random and duplicated,

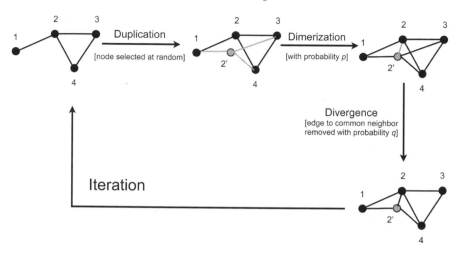

FIGURE 7.5: Network evolution steps in the DMC model. A node is chosen at random and duplicated, then dimerized with probability p. From each pair of duplicated edges, one of the edges is removed with probability q.

that is, a new node i' that links to all the neighbors of i is created. With probability p, a link between i and i' is established, representing interaction of a protein with its own copy (dimerization).

(b) *Divergence and functional complementation step*: For each common neighbor j of i and i', one of the two edges (i, j) and (i', j) is chosen at random and removed with probability q.

Note that step (b) models random functional loss due to mutation by the mechanism of edge removal, yet ensures that both duplicates taken together carry the full set of functions of the ancestral gene, modeled by the full set of edges between the two duplicate nodes and their neighbors. In this minimal model, called the duplication-mutation-complementation (DMC) model, beneficial mutations that lead to the acquisition of novel functions represented by new protein-protein interactions are not accounted for, because no new edges are created by the above rules. It has been estimated (Wagner, 2001) that the probability of *neofunctionalization*, that is, acquisition of a new edge, is relatively small for the yeast protein interaction network. Further, Vázquez et al. (2003) found that the inclusion of a small edge creation probability does not significantly change the overall structure of the resultant network.

Box 7.2 Zipf's law.

Suppose we have a collection of items (or individuals, or genes, or plants, etc.) of different types, and we rank a type according to the number of items of that type that exist in the collection. Thus, the type that has the largest frequency (f_1) of occurrence in the collection gets rank 1, the type with the second-largest frequency (f_2) of occurrence gets rank 2, and so on. If the relationship between frequency and rank follows a power law, that is, $f_n \propto n^{-\gamma}$, the collection is said to follow *Zipf's law* or a *Zipfian distribution*. For reasons that are beyond the scope of this book, Zipf's law holds for a number of disparate collections: frequencies of word occurrences in many languages, cities ranked by population, etc. Zipf's law is clearly revealed by a straight line in a log-log plot of frequency versus rank. Such a plot is called a *Zipf plot*.

In the DMC model, the parameter that controls the relative importance of the duplication versus the divergence-complementation processes is q, the probability of edge removal. For $q < 1/2$, the network dynamics is "duplication-dominated," that is, predominantly controlled by duplication, and the mean degree of the network in this case is found to grow without bound, as also occurs in other mixed duplication-mutation models (Raval, 2003). For $q > 1/2$, on the other hand, the dynamics is "divergence-dominated," and the mean degree settles to a stationary value.

Vázquez et al. (2003) simulated this model on the computer by starting with two nodes connected to each other and repeatedly applying rules (a) and (b) until they reached a network that had as many nodes as the protein interaction network of yeast. They found reasonable agreement between various topological features, including clustering coefficient and average degree, of the yeast protein interaction network and the model network with $p \simeq 0.1$ and $q \simeq 0.7$. In addition, the estimated value of p is in reasonable agreement with the proportion of self-interacting proteins in yeast (~ 0.04). Finally, Vazquez and co-workers found that Zipf connectivity plots (see Box 7.2) for the yeast network and the model network show very good agreement.

The DMC model was found to be the "best-fit" model for the *Drosophila melanogaster* protein interaction network (Middendorf et al., 2005) among seven different types of network evolutionary models culled from the literature, including a duplication-mutation scheme in which edges from the duplicated node to its neighbors can be removed

Box 7.3 Gene fission.

Gene fission is, quite simply, the process by which a single gene is split into two separate genes. In *Drosophila*, the process of gene fission appears to have occurred by duplication followed by partial degeneration of function in the two duplicates over the course of evolution in such a manner so as to preserve the full set of functions of the ancestral gene. This is exactly the type of process that is simulated in the DMC model of network evolution.

and new edges from the duplicated node to any other existing node can be created; a preferential attachment scheme; an Erdös-Renyi random network; a small-world network; and a node aging model in which the probability of a node acquiring a new edge decreases with the age of the node. The *Drosophila* protein interaction network has a relatively high clustering coefficient, a feature that is readily reproduced by the DMC mechanism because of the possibility of a node to connect to its duplicated copy. The DMC mechanism is also consistent with evidence of the occurrence of a single gene fission event (Box 7.3) in *Drosophila* 1 to 2 million years ago (Wang et al., 2004).

7.4.3   A Generalized Model of Duplication Followed by Divergence

While the DMC model incorporates functional divergence between the two members of a duplicated pair by edge deletion, the mechanism for doing so is relatively simple in the sense that a single parameter q that controls the extent of divergence dictates the probability of deletion of all duplicate edges. Furthermore, the process of evolution does not necessarily preserve the full set of biological functions of an organism, as they are in the DMC model. In recognition of these shortcomings, Evlampiev and Isambert (2008) developed a more general duplication-divergence (GDD) model for protein-protein interaction networks that includes partial genome duplications (with single gene duplications and whole genome duplications as special cases), and in which edge deletion probabilities depend on whether the two nodes connected by the edge are nonduplicated nodes, existing nodes that were duplicated, or new nodes formed as a result of duplication. Like the DMC model, no neofunctionalization occurs in the GDD model. Unlike the DMC model, dimerization is not explicitly accounted for in the GDD model, although Evlampiev and Isambert show that the inclusion of dimer-

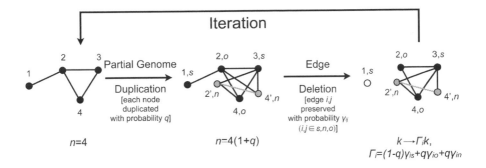

FIGURE 7.6: A schematic representation of the duplication and divergence steps in the GDD model. Each node is duplicated with probability q. Nodes in the resulting network are labeled depending upon whether they are singleton or nonduplicated nodes (s), old nodes that were duplicated (o), or newly formed nodes as a result of duplication (n). During the divergence step, an edge (i, j) is removed with probability $1 - \gamma_{ij}$, where γ_{ij} depends on the i and j node labels (i.e., on whether they are s, n, or o). The edge deletion process may result in isolated nodes (exemplified by the nonconnected unfilled circle in the final network above) that do not take part in further network dynamics. The rate of change of node degree of node i after a single duplication-divergence step is given by $\Gamma_i = (1 - q)\gamma_{is} + q\gamma_{io} + q\gamma_{in}$.

ization has little effect on network statistics after a large number of rounds of duplication-divergence.

The GDD model is schematically depicted in Figure 7.6. The model has seven basic parameters: the probability q of duplicating each node during the duplication step, and six independent probabilities of edge retention during the divergence step, namely $\gamma_{ss}, \gamma_{oo}, \gamma_{nn}, \gamma_{so}, \gamma_{sn}$, and γ_{on} ($\gamma_{sn} = \gamma_{ns}$ etc. because we assume that the network is undirected), where s, o, and n are singleton, old, and new nodes, respectively, as described in the caption of Figure 7.6. Local, partial, and whole genome duplication in the model is effected by whether $q \ll 1$, $q < 1$, or $q = 1$, respectively. The rate of increase of node degree of node i is given by Γ_i as described in Figure 7.6, with $\Gamma_i < 1$ representing the decrease in connections to node i and $\Gamma_i > 1$ representing an increase in these connections. It is assumed in the model that old duplicates lose fewer connections than new ones, that is, $\Gamma_o \geq \Gamma_n$ (this condition is not restrictive in practice because it can be used as the defining condition to distinguish between "old" versus "new" duplicated nodes). It turns out that this model displays a rich range of functional and topological features of the kind that are expected to occur in natural protein-protein interaction networks, depending upon the values that the various parameters take.

Consider, for example, the competition between two effective parameters: the edge growth rate $\Gamma = (1 - q)\Gamma_s + q\Gamma_o + q\Gamma_n$, and the rate of growth of the most conserved (i.e., older) duplicate lineage $M = (1 - q)\Gamma_s + q\Gamma_o$. By definition, $M < \Gamma$. There are three possibilities:

- $M < \Gamma < 1$. In this case, the network decreases in size and eventually vanishes. This case is therefore not biologically relevant.

- $M < 1 < \Gamma$. The network grows with time; but because $M < 1$, individual nodes (proteins) are not conserved in time and quickly drop out of the network as they lose all their connections. To keep the network growing, they are replaced by newly duplicated proteins. In this scenario, the network after a large number of duplication-divergence steps is populated mainly by newly formed proteins and retains little evolutionary memory. Proteins in a single network realization are therefore more similar to each other than proteins across different network realizations that start from the same seed network. This case, too, does not appear to be biologically relevant because of the widespread protein similarity (orthology) that is known to exist across different species.

- $1 < M < \Gamma$. This case appears to be the most biologically plausible one, because it describes growing networks in which most proteins are conserved and do not "drop out" of the network. A snapshot of such a network at late times should contain a large proportion of old proteins in its core, corresponding roughly to the empirical evidence[6].

Given that $1 < M < \Gamma$ appears to be a constraint dictated by the biological observation of orthology across species, we can explore the different types of network dynamics that are possible under this constraint. Evlampiev and Isambert (2008) show that another effective parameter controls the topology of the evolved network. This parameter is $M' = \max_i(\Gamma_i)$, which is the largest rate of edge growth among

---

[6]Note that the case of pure duplication in the language of the GDD model corresponds to $\gamma_{ij} = 1$ for all i, j. In this limit, $\Gamma_i = 1 + q$ for all i and $M = 1 + q > 1$ Thus, as expected, the pure duplication model also retains evolutionary memory.

the three different node types. Note that, because M is a weighted average of two edge growth rates, it must be true that $M < M'$. It is found that $M' < 1$ (which necessarily corresponds to the biologically uninteresting case $M < 1$) leads to networks whose degree distribution decays exponentially, $p_k \propto e^{-\mu k}$, while for $M' > 1$ a number of interesting behaviors are possible. In summary, some of the salient findings are

- In the local duplication-divergence limit ($q \ll 1$ and $\gamma_{ss} = 1$, corresponding to complete retention of edges connecting non-duplicated nodes), and when γ_{so} is close to 1, the network approaches a power-law degree distribution with exponent between 1 and 3, depending upon detailed parameter values.

- In the whole-genome duplication-divergence limit ($q = 1$), the network approaches an exponentially decaying degree distribution if $M < 1$ (with $M' > 1$). If $M > 1$, a number of regimes of power-law distributions are possible if $\gamma_{oo} \neq \gamma_{nn}$ (asymmetric divergence between old and new duplicates). However, if $M > 1$ and $\gamma_{oo} = \gamma_{nn}$, one obtains nonstationary solutions that are highly dense networks whose mean degree grows as the network grows. Such networks do not appear to be biologically relevant.

- In general, if the seven fundamental parameters are allowed to vary over the course of evolution, networks with $M > 1$ always have power-law distributions (or are dense) in the GDD model, while networks with exponentially decaying distributions always have $M \leq M' < 1$.

- For $M > 1$, although individual proteins are conserved, network motifs containing two or more proteins are not conserved. Thus, sequence or structure similarity at the level of individual proteins in different organisms does not necessarily translate to functional similarity, if function is defined in terms of the set of interactions of the protein.

- The GDD model can be fitted well to the the yeast protein-protein interaction network in terms of both the standard degree distribution as well as the distribution of average degree of the first neighbor of a randomly chosen node.

7.5 Reconstruction of Ancient Networks from Modern Ones

It is important to keep in mind that the fundamental timesteps in the network evolution models that we have studied in reality represent huge leaps in time because events such as duplications and mutations that are fixed in the population occur over evolutionary timescales, and not every such event leads to rewiring of the network. For example, the average protein-protein interaction network rewiring rate since the time of species divergence between the human and the fruit fly (approximately 800 million years ago) is estimated at 10^{-6} per edge per million years (Shou et al., 2011). Thus, in a network comprising about 100,000 interactions, a single edge would be rewired every 10 million years on average. Being able to infer what networks looked like many timesteps back would therefore allow us to reconstruct very ancient events.

Reconstruction of ancient networks from modern ones is also an acid test of the network evolution model, as the reconstructed data can be confronted with phylogenetic estimates of the ages of proteins. It is in general a very difficult computational problem, but significant advances that were aided by a few simplifying assumptions were made in the work of Navlakha and Kingsford (2011).

The basic methodology of network reconstruction, given a network evolution model $\mathcal{M}$, proceeds as follows. Suppose we have a snapshot of the network G_t at time t and would like to infer the network at time $t - \Delta t$. A natural procedure is to find the most probable network at time $t - \Delta t$ given $\mathcal{M}$ and the network at time t:

$$G_{t-\Delta t}^* = \underset{G_{t-\Delta t}}{\operatorname{argmax}} P(G_{t-\Delta t} \mid G_t, \mathcal{M}), \qquad (7.36)$$

where $P(\cdot)$ is the probability of the ancient network given the present one, and the "argmax" operation corresponds to finding the most probable network at time $t - \Delta t$ given the network at time t.

Typically, the space of ancient networks grows exponentially as we try to infer networks further back in time, that is, with increasing Δt. It is therefore not feasible to search this space and directly compute $G_{t-\Delta t}^*$ for an arbitrary Δt. A more feasible procedure is to compute the most probable network only for a single backward timestep of the evolutionary model, that is, $\Delta t = 1$, and then assume that the most probable network at timestep $t - 2$ can be reconstructed from the most probable network at timestep $t - 1$, and so on. This simplifying assumption avoids the exponential blow-up in the search space as one proceeds backward in time. Equation (7.36) with $\Delta t = 1$ can be expressed in a more tractable form using Bayes' rule:

$$
\begin{aligned}
G^*_{t-1} &= \operatorname*{argmax}_{G_{t-1}} P(G_{t-1} \mid G_t, \mathcal{M}), \\
&= \operatorname*{argmax}_{G_{t-1}} \frac{P(G_t \mid G_{t-1}, \mathcal{M}) P(G_{t-1} \mid \mathcal{M})}{P(G_t \mid \mathcal{M})}, \\
&= \operatorname*{argmax}_{G_{t-1}} P(G_t \mid G_{t-1}, \mathcal{M}) P(G_{t-1} \mid \mathcal{M}), \qquad (7.37)
\end{aligned}
$$

where the last equality follows because the "argmax" operation runs only over networks at time $t-1$, not networks at time t. A further simplification avoids the problem of computing the "prior" probability $P(G_{t-1} \mid \mathcal{M})$ by assuming that this probability is the same for all networks that can be realized by the model $\mathcal{M}$. In reality, the problem of finding frequencies of occurrence of every possible network of a certain size that can be realized by a network dynamics model from any initial seed network is not easily solved.

With the additional simplification mentioned above, the most probable reconstructed network one timestep back can be expressed entirely in terms of forward probabilities:

$$
G^*_{t-1} = \operatorname*{argmax}_{G_{t-1}} P(G_t \mid G_{t-1}, \mathcal{M}). \qquad (7.38)
$$

Note that the forward probability $P(G_t \mid G_{t-1}, \mathcal{M})$ is readily computable from the dynamics of the network model under consideration.

7.5.1 Network Reconstruction Based on the DMC Model

In the previous section, we discussed the DMC model where a network grows by node duplication followed by possible dimerization and divergence. Using this model as an example, we now outline the steps in the computation of the forward probability $P(G_t \mid G_{t-1}, \mathcal{M})$ so that we can reconstruct the network one timestep back. Assuming that some, as yet unknown, node was duplicated at time t, let us consider arbitrary nodes u and v in the network G_t and assume that v is a newly created duplicate of u. Then the appropriate forward probability can be written as

$$
P(G_t \mid G_{t-1}, \mathcal{M}) = \frac{p_{uv}}{n} \prod_{N(u) \cap N(v)} (1-q) \prod_{N(u) \triangle N(v)} \frac{q}{2}, \qquad (7.39)
$$

where $p_{uv} = p$ if u and v are directly connected (i.e., dimerization has occurred) and $p_{uv} = 1 - p$ otherwise. The set $N(u) \cap N(v)$ represents the set of common neighbors of u and v—these are interactions that have not been deleted, a phenomenon which occurs in the DMC model with probability $1 - q$ for each edge. The set $N(u) \triangle N(v)$ represents the set of nodes that are either neighbors of u or of v but not both—these are interactions in which one of the duplicate edges has been lost in the divergence step, a phenomenon that occurs with probability $q/2$ for each edge (the factor of $1/2$ occurs because only one of the duplicate edges is selected at random and removed).

Network reconstruction is carried out by computing the right-hand side of Equation (7.39) exhaustively for all pairs of nodes u, v in the network G_t and finding the pair that maximizes $P(G_t \mid G_{t-1}, \mathcal{M})$. Once this pair is found, the edges that connect to u and the edges that connect to v are together assigned to u (these edges would have connected to u before divergence occurred), and the node v is deleted. This process yields the reconstructed network one timestep back. The entire process can be iterated to reconstruct networks in the distant past.

A number of interesting results arise in the reconstruction of ancient networks from present ones using the DMC and other models (Navlakha and Kingsford, 2011). As mentioned above, one of the predictions from network reconstruction is the relative ages of various nodes in the network. Using the present-day yeast protein interaction network as a starting point for backward reconstruction, Navlakha and Kingsford showed that the DMC model best reproduced relative ages of proteins as compared to other models that are more appropriate for the description of social networks, like preferential attachment. The DMC model parameters that provided the best prediction of relative ages correspond to high p and low q, in broad agreement with a variety of data sources. Pairs of duplicates identified by the reconstruction procedure also shared significantly higher average sequence identity than expected by chance (although sequence information was not used to guide the reconstruction), and were found to be likely to participate in the same protein complex.

7.6 Large-Scale Biophysics-Inspired Models for Protein Interaction Networks

The large-scale network models we have considered so far fall into two main classes: static, "non-biological" random models (Erdös-Renyi, configuration, and small-world models) and dynamical models that attempt to distill basic principles of social or biological evolution into rules for network modification (preferential attachment-based and duplication-based models).

A parallel line of investigation makes use of the biophysical description of molecular interactions to build networks rather than relying on general growth principles based on the evolution of genomes. One of the most important models of protein-protein interactions that represents this area of network biology research is the so-called MpK model by Deeds et al. (2006). This model was shown to reproduce topological features of high-throughput protein interaction network data in yeast. As discussed below, the MpK model also serves as an important null model against which real protein interaction networks can be compared. It was one of the first models to directly use the hydrophobicity of protein surface patches in large-scale binding propensity calculations. Two other models that we outline here use abstracted versions of this idea in interesting ways.

7.6.1 The MpK Model

The motivation for the MpK model rests on two principal observations. First, as we already know, the yeast two-hybrid (Y2H) experiment for measuring protein-protein interactions in a high-throughput fashion is rather inaccurate and leads to "noisy" interaction networks. For example, when the results of two major Y2H screens in yeast (Uetz et al., 2000; Ito et al., 2001) are compared, only about 150 of the thousands of interactions identified in each experiment are recovered in the other experiment (Deane et al., 2002). Further, as shown by Deeds et al. (2006), the degree of a protein node in one experiment is only very weakly correlated with the degree of the *same* protein in the other experiment[7]. Even so, both sets of experimental data lead to a power-law degree distribution, with similar exponents of about 2.

---

[7]Similar disagreements exist in the high-throughput protein interaction networks of *Drosophila melanogaster* (Formstecher et al., 2005).

The second observation is that two different Y2H experiments on the same set of proteins will generally give different results because of a *nonspecific* component of the process of protein-protein interaction: proteins that contain a large number of surface hydrophobic residues will tend to have a greater propensity to bind with other proteins along the portion of the surface that contains these hydrophobic residues, *regardless* of the nature of the other protein. Note that hydrophobic residues should generally not occur on a portion of protein surface that is exposed to water, and therefore there is a greater tendency for such surfaces to function as binding interfaces to other proteins. The water molecules will elute from such a binding interface, thus stabilizing it.

Deeds et al. (2006) posit that this nonspecific binding is essentially sufficient to explain both the disagreement between independent Y2H experiments on the same set of proteins as well as their agreement with regard to the power-law degree distribution and some other topological characteristics.

The MpK model fundamentally involves three parameters M, p, and K that are computed for each protein in the model. For a given protein, p is the proportion of amino acid residues residing on the protein surface that is hydrophobic, M is the number of exposed residues that are involved in binding to other proteins, and K is the number of residues counted in M that are hydrophobic. Generally, M is not the total number of exposed residues as not all exposed residues will be involved in binding. Given p and M, the probability of finding K hydrophobic residues within the set of M exposed residues involved in binding follows a binomial distribution:

$$p(K) = \binom{M}{K} p^K (1 - p)^{M-K}. \tag{7.40}$$

Thus, given p and M for an individual protein, we can sample values of K for it from the above distribution. We discuss below how p and M can be estimated for each protein. For now, let us assume that p and M are known for each protein, and that therefore a (random) value of K can be assigned to the protein according to the binomial distribution above.

Next, the model assumes that the free energy F_{ij} of binding of proteins i and j is additive in the total number of hydrophobic residues available for binding. In appropriate units, therefore, F_{ij} is simply minus the sum of the K values, that is, $F_{ij} = -(K_i + K_j)$. Thus, two proteins with a large total number of hydrophobic exposed residues

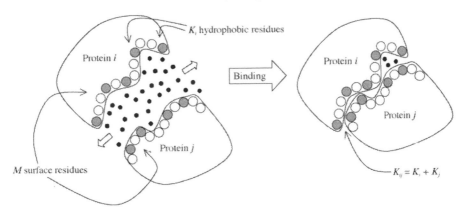

FIGURE 7.7: The MpK model for protein interaction networks. Binding between two proteins results in the burial of hydrophobic groups and is therefore determined by the number of exposed hydrophobic residues. Reproduced with permission from Deeds et al. (2006). [©2006 National Academy of Sciences, U.S.A.]

that are involved in binding will tend to have a low free energy of binding. This is because the binding of these two proteins will result in the thermodynamically favorable burial of a large number of hydrophobic groups: $K_{ij} = K_i + K_j$ (Figure 7.7). The actual binding affinity A_{ij} between the two proteins is given by statistical mechanical considerations as the exponential of minus the free energy:

$$A_{ij} \sim e^{-F_{ij}} = e^{K_i + K_j}. \tag{7.41}$$

The final ingredient of the model is to specify a cutoff affinity value A_C such that only protein pairs with $A_{ij} \geq A_C$ will be detected in an experiment as interacting proteins. The cutoff therefore models the sensitivity of the experiment.

To recapitulate the basic features of the model, if p and M are known for each protein, one can sample values of K for each protein and use them to find binding affinities A_{ij} for each pair of proteins. A model network is constructed by assigning an interaction edge to the pair (i, j) if the binding affinity of the pair exceeds the chosen cutoff value. A final modification to the model network consists of adding random noise by linking a number of isolated nodes in the network to other randomly chosen nodes in such a manner that the total number of nonisolated nodes in the model network agrees with that found in the real yeast network.

Properties of the resulting model network can then be studied and be compared to those of real protein interaction networks. Note that, apart from the final noise addition step, the MpK model has an inherently random feature because different realizations of the model network will differ in the K values that are sampled, even if the two realizations have the same set of p and M values. Thus, different realizations may be said to model different Y2H experiments.

We now turn to the issue of estimating the parameters p and M, and the cutoffaffi nity A_C. Deeds et al. (2006) assume that all proteins in the model have the same value of M. They also find that there is a trade-off between the value of M and the value of A_C: any change in M can be compensated for by a corresponding change in A_C. For example, increasing M would increase the number of exposed residues available for binding and thus sytematically increase sampled values of K and the pairwise binding affinities. This can be compensated for by making the affinity cutoff more stringent (increasing A_C), thus allowing fewer pairs to be modeled as "interacting" proteins. Because of this trade-off, Deeds et al. kept the value of M fixed at $M = 100$ in all their simulations.

The value of p for each yeast protein is found by calculating the proportion of exposed residues that are hydrophobic from the three-dimensional structure of the protein, or, when the structure is unknown, inferring this proportion by homology to other proteins of known structure. These values taken together are found to be normally distributed to good approximation. Values of p are then sampled from this normal distribution and used in the model. Thus, the only remaining free parameter in the model is the cutoffaffi nity A_C.

It is found that the MpK model generates networks with power-law degree distributions for values of A_C in the range of $\sim 10^{25}$ to 10^{40} with values of the power-law exponent increasing as A_C is increased, and ranging from ~ 1.25 to 3.5. In particular, the exponent of about 2 found in the Y2H experiments in yeast is well modeled by an appropriate choice of A_C. Figure 7.8 shows a comparison of the model network degree distribution with its Y2H counterpart. The model is also found to correctly predict the heirarchical structure of the Y2H network, defined as the power-law scaling of clustering coefficient with degree. It further makes a number of testable predictions regarding the correlation between protein connectivity and its surface hydrophobicity, and suggests that the data found in high-throughput protein interac-

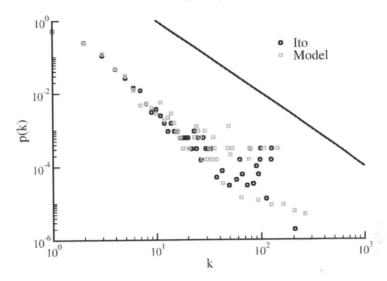

FIGURE 7.8: Comparison of degree distributions of the yeast protein inter-action network [from Ito et al. (2001)] and a realization of the MpK model network. The straight line represents a power law with an exponent of 2. Re-produced with permission from Deeds et al. (2006). [©2006 National Academy of Sciences, U.S.A.]

tion studies is largely dominated by nonspecific interactions arising out of the inherent "stickiness" of a protein attributed to its surface hy-drophobicity. This observation clearly shows that results from single high-throughput protein interaction experiments cannot be used reli-ably but rather in conjunction with other data sources to produce "high confidence" networks [see, for example, Han et al. (2004); Reguly et al. (2006)], as we have discussed in Chapter 5. Finally, it may sometimes be more appropriate to assess the significance of topological features found in real protein interaction networks by comparing them to an *MpK*-like biophysics-based model networks rather than configuration model networks, because the former incorporates features that arise out of nonspecificity of real protein interactions[8]. We now discuss another model based on protein stickiness that underscores this idea.

---

[8]Note, however, that these types of models only make sense for physical interaction networks but not for the more abstract metabolic and genetic interactions. For the latter networks, the configuration model and its variants continue to be useful null models.

7.6.2 A Configuration-Like Model Based on Protein "Stickiness"

The idea of using surface hydrophobicity information to determine nonspecific binding propensity and therefore to model nonspecific interaction networks can be turned around to ask to what extent real protein interaction networks can be described by a combination of nonspecific binding propensities or "stickiness" indices (Przulj and Higham, 2006). This type of construction is an alternative to the standard configuration model. Here, as opposed to directly using the degree sequence of a real network to generate a random network model, a set of stickiness indices derived from the degree sequence is used. The basic assumptions in this formulation are: (1) proteins of high degree have a large propensity for nonspecific binding to other proteins—thus the binding propensity (or stickiness index) of a protein is a nondecreasing function of its degree; and (2) the probability of interaction of two proteins is proportional to the product of their stickiness indices.

Given the two assumptions above, combined with the constraint that the mean degree of protein i in the model network should be equal to its degree in the real interaction network, it can be shown (Przulj and Higham, 2006) that the stickiness index of protein i can be uniquely represented as

$$\theta_i = \frac{k_i}{\sqrt{\sum_j k_j}}, \tag{7.42}$$

where k_i is the degree of protein i in the real network. The algorithm for constructing the stickiness-based random network is as follows. First, we compute stickiness indices for all nodes in the network using the degree information. Then, given two nodes i and j, we connect them with probability $\theta_i \theta_j$. (On the computer, this is achieved by generating a random number uniformly between 0 and 1. If this number is less than $\theta_i \theta_j$, we connect the nodes i and j by an edge; otherwise, we do not connect them). This process is repeated for all pairs of nodes in the network.

It is found that random networks generated using the stickiness model mimic features of real protein-protein interaction networks to higher accuracy than standard configuration models and preferential attachment models. Like the configuration model, the stickiness framework has no additional parameters once the degree sequence is specified. This finding suggests that the topological structure of protein-protein interaction networks has a strong biophysical basis.

7.6.3 Growing Crystals on Evolutionary Timescales

The biophysical notion that the limited availability of protein binding surface restricts binding to multiple proteins is the basis of an edge connection mechanism that may be termed "Anti-Preferential attachment" (AP), in which a new node preferentially attaches to nodes that have few connections as opposed to many connections. While this mechanism does not, by itself, produce networks that have features in common with real protein-protein interaction networks, it serves as a biophysical basis for a dynamical network model in which AP is combined with the propensity of proteins to form complexes over the course of evolution. This model is aptly termed the Crystal Growth (CG) model because the network growth dynamics in this model is similar to that of crystals growing in solution (Kim and Marcotte, 2008), albeit the latter growth occurs over far shorter timescales than the former! The model uses the concept of *modules*, which we first encountered in Chapter 5. Modules are approximately defined as densely connected regions of the network whose nodes have far more intra-module connections than inter-module connections. There are various ways to define modules more precisely, as we shall see in the next chapter. Here, it is sufficient to retain the qualitative notion that a module is a densely connected sub-network in a much larger network.

In the CG model, at each timestep a new node with a fixed number m of edges is introduced. With some small probability p_{new}, the node becomes a seed for a new module and connects in accordance with the AP rule to existing nodes in the network. However, with a much larger probability $1 - p_{new}$, the node picks an existing module at random, connects one of its edges to an existing node in the module in accordance with AP, and connects the rest of its edges only to neighbors of the existing node to which it is already connected (these are typically, but not always, other nodes in the *same* module). If protein interaction modules are thought of as crystals, it is easy to imagine that these dynamics mimic the growth of crystals in solution. Over evolutionary timescales, they model the growth and origin of protein complexes.

When proteins are classified into different age groups and these groups are mapped onto the experimentally found protein-protein interaction networks in yeast, a remarkable age dependency of interactions is observed: proteins are more likely to connect to close ancestors rather than distant ancestors. Perhaps equally remarkable, standard preferential attachment and duplication-divergence models do not reproduce this feature. A pure AP model does reproduce the observed age dependency but fails to reproduce the correct degree distributions and clustering behavior. Among the set of models studied by Kim and Marcotte (2008), the CG model was unique in its ability to reproduce topological features of real protein-protein interaction networks as well as the age dependency of interactions. It also accurately reproduces two other features of yeast protein interaction networks, namely that members of the same protein complex are likely to belong to similar age groups, and that homodimers (proteins that interact with copies of themselves) occur with higher frequency among older proteins than younger ones.

Chapter 8

Network Modularity and Robustness

In previous chapters of this book, our emphasis has been mainly on modeling biomolecular networks by experimental, theoretical and computational means, and on the validation and filtering of these models. We have, to a rather limited extent, discussed the use of network information for the inference of biological function and to help clarify disease states. Furthermore, we have barely touched upon how knowledge of interactions and of their evolution can help illuminate deeper issues such as the modularity of biological organization and organismal robustness. These questions are addressed in greater depth in this chapter and the one following it. We are now in a position to ask how network data can be systematically used to advance biology and medicine.

A key concept central to the usefulness of the network approach to biology or medicine is that of modularity. The modular organization of biological networks has received much attention in recent years, primarily because the modularity of connection patterns is intimately connected to the modularity of biological function. The complexity of biological function can best be understood in terms of components that retain some of the biological properties of interest. Hence it is desirable to reduce complexity down to structural or functional units: the modules. Understanding modularity at the network level allows one to form a basis for understanding the evolution of modularity at the functional level, the links between different functions, and the correlation of their phenotypes. Furthermore, the existence of network modules that are densely interconnected among themselves while being sparsely connected to the rest of the network hints at a type of robustness: targeting a component of a module is likely to degrade the function(s) associated with the specific module while not seriously affecting other modules or the whole organism. We therefore study the topics of modularity and robustness, as they manifest themselves at the network level, together in this chapter. We begin with a classic example in biology: segmental patterning in the embryonic stage of development of the fruit fly

Drosophila melanogaster. Following this, we discuss definitions of modularity in larger networks and their functional significance. Finally, we address dynamical and topological robustness in networks.

8.1   Modularity Implies Robustness in the *Drosophila* Segment Polarity Network

The outer layer of cells in an organized tissue is called the *epidermis*. Pattern formation in the epidermis of the *Drosophila* embryo is often considered one of the prime examples of developmental organization that is amenable to detailed study at the molecular level. Like all insects (and, more generally, *arthropods*: invertebrate animals that have a segmented body and jointed appendages), the fruit fly body is segmented. It is composed of fourteen segments: three that make up the head, including antennae and mouth, three that make up the thorax—the anterior portion of the body (each thoracic segment supports a pair of legs), and eight abdominal segments. The process of segmentation begins very early in development; the body plan is laid by the formation of gradients of maternally derived mRNA molecules (made from a set of *maternal effect* genes) along the *Drosophila* egg. Following fertilization of the haploid egg by the haploid sperm, the process of nuclear division of the diploid cell or zygote (mitosis) begins. However, nuclear division is not accompanied by division of the cytoplasm (called *cytokinesis*). Instead, the early *Drosophila* embryo is a multinucleate cell, called a *syncytium*, which consists of a swarm of nuclei floating in a shared cytoplasm. These nuclei migrate to the periphery of the embryo at the tenth nuclear division; at the thirteenth division they are partitioned into separate cells all located at the periphery. This is the *cellular blastoderm*[1].

The peripheral cells in the blastoderm encounter different levels of certain crucial maternal mRNA molecules, depending on the location of the cells. The cells, in response to these maternal mRNAs, show distinct periodic patterns of mRNA and protein expression of their own that govern the eventual differentiation into physiological segments in the adult fly. Thus, the process of segmentation begins well before the egg

---

[1]For a detailed description, the reader should refer to a modern developmental biology textbook, such as Gilbert (2000).

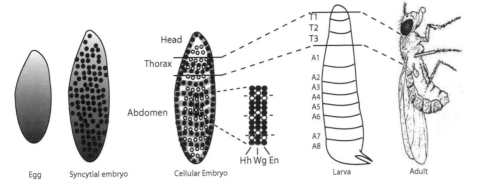

FIGURE 8.1: Stages in *Drosophila* development from egg to fully matured adult. The egg cytoplasm contains gradients of maternal mRNA concentration established along both axes: anteroposterior and dorsoventral. The dark circles represent nuclei. The inset in the middle of the figure shows two successive parasegments (whose boundaries are set off by horizontal lines), with the nuclei expressing Hh, Wg, or En as indicated by the filled circles (an empty circle signals the absence of expression of the corresponding gene). Thus the topmost cell in the inset expresses Hh, Wg, and En; the second cell expresses Hh and En; and so on. These combinatorial expression patterns, coupled with the positional information provided by the maternal mRNA gradient, are responsible for setting up the segment identity of the larva, and ultimately of the adult fly.

is fertilized, and is initiated by the expression gradients of the *maternal effect* genes: most importantly, the *bicoid* and *hunchback* genes whose mRNA and protein gradients determine the patterning of the head and thorax (the anterior portion of the body), and the *nanos* and *caudal* genes whose expression gradients determine the eventual segmentation of the abdomen (the posterior portion of the body). Note that these gradients are established while the embryo is still a syncytium (Figure 8.1).

The expression gradients of the maternal effect genes trigger the activity and expression gradient formation of the *segmentation genes* (whose mRNA molecules are transcribed from embryo cells, not the maternal cells), which, in turn, consist of three major types that are activated in sequence: the *gap* genes, the *pair-rule* genes, and the *segment polarity* genes. Broadly speaking, the expression gradients of the gap genes subdivide the syncytium into major segments, those of the pair-rule genes establish periodic pairs of segments at the blastoderm

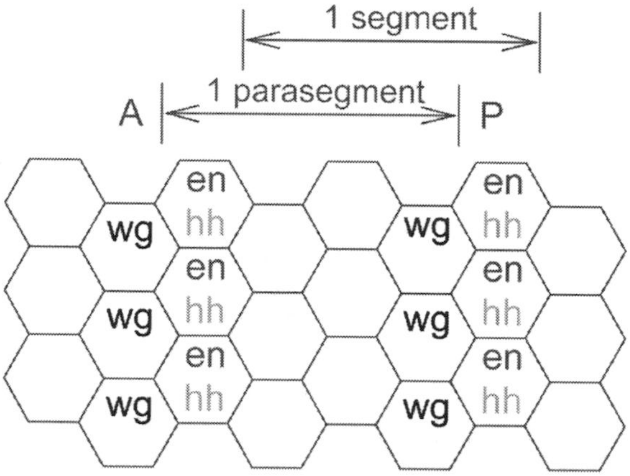

FIGURE 8.2: Expression of the main segment polarity genes and delineation of parasegments and segments in *Drosophila*. Hexagons denote individual cells, and gene labels within a cell indicate the genes that are dominantly expressed within that cell. The expression pattern shown is the one that is stably maintained by segment polarity genes in the wild-type fruit fly. Note that parasegments are the boundaries across which diffusion of the secreted signaling proteins, namely Hh and Wg, occurs. Anatomical segments in the adult fly are offset from the parasegments. "A" and "P" denote Anterior and Posterior positions along the fly body, respectively. Adapted with permission from Ingolia (2004).

stage, and those of the segment polarity genes establish the antero-posterior orientation within each segment. Taken together, the expression gradients of the segmentation genes along the cells comprising the blastoderm specify fourteen *parasegments* that are closely related to the final anatomical segments (Figures 8.2 and 8.3). Further down the cascade, the segmentation genes regulate yet another family of development control genes, the *homeotic selector* genes. The genes involved in early *Drosophila* development thus form a transcriptional cascade, with each set of genes effecting the transcription of the next set of genes at a later point in time.

Our focus here is on the maintenance of the expression pattern of the segment polarity genes. While most of the genes mentioned above are activated in a transient fashion, stable pattern formation in the form of parasegments is maintained by the segment polarity genes via their interactions across cells belonging to adjacent parasegments. The segment polarity genes are also strongly evolutionarily conserved: their homologs have been identified even in vertebrates, including humans. In vertebrates, these genes, among other roles, are crucial for laying down the patterns of brain and spinal cord segmentation. Three segment polarity genes play a distinctive role in the maintenance of parasegments: the *wingless* (*wg*), the *engrailed* (*en*), and the *hedgehog* (*hh*) genes[2]. A stable parasegment in the blastoderm is a row of four cells, with the cell at the anterior position expressing *en* and *hh*, the middle two cells not expressing any of the three genes, and the cell at the posterior position expressing *wg*. The four cells also display characteristic expression patterns of the other segment polarity genes, but we do not discuss those here. As shown in Figure 8.2, segment boundaries between the anatomical segments in the adult fly are offset from the parasegment boundaries.

Regulatory interactions among the segment polarity genes can be combined to construct a core segment polarity network (Figure 8.3), which is a small network responsible for the maintenance of the parasegmental expression pattern of *wg*, *en*, and *hh*. While the protein products of *wg* and *hh* are secreted outside the cell, the protein corresponding to *en* is a transcription factor responsible for the transcription and regulation of the gene *hh*. As is clear from Figure 8.3, the diffusion of the Wg protein from a cell can maintain a high level of En (and therefore of Hh) in a neighboring cell, and conversely, the diffusion of Hh from a neighbor, via reduction of expression of the Cn protein, can maintain a high level of Wg.

The core *Drosophila* network is also consistent with the observed low expression of *wg*, *en*, and *hh* genes in half the cells of the parasegment. The reason that *wg*, *en*, and *hh* can each exist in two expression states depending on the cell (one where they have high expression and the other where they have negligible expression) can be traced to the direct auto-activation of *wg* and that of *en* (via the inhibition of the *sloppy paired gene, slp*). While we do not discuss the details here, it is well known that systems with auto-activation (or positive feedback) exhibit *bistability*, that is, such systems can exist in any one of two different

---

[2]Like many other genes, the names of genes involved in *Drosophila* development are based on the phenotype of the organism that results from a mutation of the corresponding gene. Thus, loss of function mutation of the *wg* gene leads to wingless flies, and loss of function mutation of the *hh* gene leads to fly larvae that have a hedgehog-like appearance.

stable steady states, with one of the two steady states corresponding to low activity and the other one to high activity [see, for example, Ferrell (2002); Smits et al. (2006)]. The eventual adoption of the final state depends on the initial conditions of the system. Thus, auto-activation leads to bistability, which in turn allows for the existence of different patterns of expression among the four cells of the parasegment.

The functional roles of components of the core segment polarity network can be abstracted to form the simplified network shown in the lower panel of Figure 8.3. This simplification is possible because we are mainly interested in the steady-state behavior of the core network and not on the transient expression of the segment polarity genes. First, we can clearly combine all mRNA nodes with their protein counterparts because the core network does not involve post-transcriptional regulation of the mRNA. Next, all the *hedgehog* and *engrailed* nodes can be merged because the expression of *hedgehog* is a monotonic function of that of *engrailed* and depends solely upon it. Thus, a single node labeled "E" denotes the state of both *engrailed* and *hedgehog* in the simplified network. Furthermore, since the action of extracellular *Hh* is to ultimately activate *wg* and the action of extracellular *Wg* is to activate *en*, these activations are shown as direct but dashed activated interactions in the simplified network. Lastly, the interactions between "E" and the node "S" representing the mRNA and protein corresponding to the *slp* gene are repressing and are represented as such—as pointed out above, the two repressing interactions lead to bistability in "E" and "S."

Every regulatory interaction in the segment polarity network can be modeled in terms of a differential equation describing the rate of formation of each entity, similar to the reactions that we encountered in Chapter 6. While the mathematical forms of these differential equations are not important here, it is important to note that the equations contain parameters whose values must be known in order to model or simulate the segment polarity network. These parameters include rate constants that specify the speed of transcription, diffusion constants that specify the rate at which secreted proteins diffuse from one cell to the next, and Michaelis-Menten parameters (see Box 8.1) that specify the binding affinity between enzyme and substrate. The segment polarity network contains nearly fifty such parameters that must be specified for modeling it. While the values of some of these parameters are known from *in vitro* experiments, the values of most of the parameters are unknown.

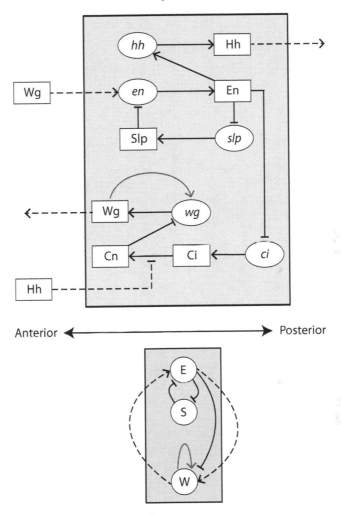

FIGURE 8.3: The core segment polarity network in *Drosophila* (Ma et al., 2006). Ellipses represent mRNAs and rectangles represent proteins. Lines ending with an arrow and a bar denote activation and repression, respectively. Dashed lines indicate proteins secreted by the cell, which diffuse to neighboring cells. The auto-activation of *wingless*, denoted by a gray line, is not supported by direct biological evidence but is consistent with the maintenance of a stable stationary expression pattern in the parasegments. The displayed regulatory interactions imply that elevated levels of Hh in a neighboring cell can lead to elevated levels of Wg protein, while elevated levels of Wg in a neighboring cell can lead to elevated levels of En (and therefore of Hh) protein. Other than the genes encountered in the text, this network introduces the following: the *sloppy paired* gene (*slp*), which is a pair-rule gene that is activated before the segment polarity genes but expressed constitutively thereafter; the *cubitus interruptus* gene (*ci*); and a fragment of *ci*'s protein product, Cn, that represses the activation of *wg*. The bottom diagram is a simplified version of the top diagram, where E represents En, S represents Slp, and W represents Wg.

Box 8.1 The Michaelis-Menten equation.

The *Michaelis-Menten equation* describes the rate of formation of a chemical species whose production is catalyzed by an enzyme. Consider an enzyme E binding to a substrate S to form an enzyme-substrate complex ES that is converted to free enzyme plus a product P. This reaction may be represented as

$$E + S \underset{k_r}{\overset{k_f}{\rightleftharpoons}} ES \overset{k_c}{\longrightarrow} E + P,$$

where k_f and k_r are rate constants for the association and disassociation of the complex, respectively, and k_c is the rate constant for the formation of the product from the complex. The rates of formation of complex and product are thus given by

$$\frac{d[ES]}{dt} = k_f[E][S] - (k_r + k_c)[ES], \tag{8.1}$$

$$\frac{d[P]}{dt} = k_c[ES], \tag{8.2}$$

where $[E]$, $[S]$, and $[ES]$ are the concentrations of enzyme, substrate, and complex, respectively, and $[P]$ is the concentration of product. Because the enzyme acts as a catalyst and is not depleted over the course of the reaction, it must be that the total concentration of enzyme (free + bound to substrate) remains constant. Thus, we have the relation $[E] + [ES] = [E]_0$, where $[E]_0$ is the initial enzyme concentration.

Furthermore, we may assume the existence of a *quasi-steady state*, in which both free and bound forms of the enzyme are *separately* constant, and the only change occurring is that the amount of substrate is depleted while the amount of product increases. In this quasi-steady state, $d[ES]/dt = 0$, yielding [from Equation (8.1)] $[ES] = k_f[E][S]/(k_r + k_c)$. The further substitution $[E] = [E]_0 - [ES]$ leads to the following relation:

$$[ES] = \frac{[E]_0[S]}{K_m + [S]},$$

where $K_m = (k_r + k_c)/k_f$. Substituting the above expression back into Equation (8.2) yields

$$\frac{d[P]}{dt} = \frac{V_{\max}[S]}{K_m + [S]},$$

where $V_{\max} = k_c[E]_0$. The above equation is called the Michaelis-Menten equation, and V_m and K_m are the Michaelis-Menten parameters. It shows that the rate of formation of a product in an enzyme-catalyzed reaction increases as a function of the substrate concentration up to a maximum rate $V_{\max}$ that is limited by the enzyme concentration.

Remarkably, the steady-state behavior of the segment polarity network model is insensitive to the values of the underlying parameters. In a detailed robustness analysis, von Dassow et al. (2000) searched for parameter sets that maintained the expression pattern of segment polarity genes in parasegments. Starting from a set of parameters that maintained this pattern, they individually varied each parameter in turn and found that the steady-state expression pattern was insensitive to tenfold or more variation in the values of most parameters. Furthermore, they found that a randomly chosen set of parameters maintains the correct expression pattern with a probability of 0.5%. While this number may appear to be low, it is actually very large considering that the parameter space is a very high-dimensional space; at the level of a single parameter, it corresponds to a probability of 90% that a randomly chosen single parameter value is capable of maintaining the correct expression pattern, because $0.9^{50} \simeq 0.005$. This remarkable robustness to parameter values in fact allows one to model the network in a "parameter-free" manner: Albert and Othmer (2003) modeled the segment polarity network as a Boolean regulatory network in which each entity is either turned ON (i.e., has high expression) or OFF (low expression) and the interactions between entities are governed by simple logical relationships rather than by differential equations with parameters. They essentially found that the Boolean segment polarity network correctly models parasegmental expression and also predicts the effect of mutations of various segment polarity genes. Furthermore, it turns out that the segment polarity network is robust to the choice of initial conditions. While the function of the segment polarity network is to *maintain* the correct parasegmental expression pattern and not *create* it, in practice the starting pattern can be significantly perturbed without destroying the final steady state (von Dassow et al., 2000).

What is the cause of the impressive robustness of the segment polarity network? By comparing the steady states of networks with and without auto-activation of *wg* and *en*, Ingolia (2004) showed that this robust behavior is sufficiently explained by the occurrence of bistability. This question was investigated exhaustively by Ma et al. (2006) using the simplified network shown in the lower panel of Figure 8.3. They analyzed the robustness properties of all possible topologies involving the three meta-nodes shown there, and unearthed the topological features common to robust networks. Because every meta-node in the simplified network may regulate itself and two other nodes, both intracellularly and intercellularly, $3 \times 3 \times 2 = 18$ links are possible. Each link can signify positive regulation, negative regulation, or be absent, resulting in three possibilities for each link. Thus, the total number of possible networks involving three meta-nodes is $3^{18} = 387{,}420{,}489$, a very large number! Ma et al. (2006) narrowed their analysis to topologies in which at most two of the three nodes can signal outside the cell, thus restricting the number of possible topologies to $3^{15} = 14{,}348{,}907$. These topologies include networks in which the "S" meta-node is not connected to any other meta-node and is therefore effectively absent from the network. For each of the more than 14 million networks, they measured robustness by choosing 100 parameter sets at random and finding the proportion of these that could maintain correct parasegmental expression. If a specific topology could maintain the required expression pattern for more than 10% of the chosen parameter sets, it was considered a robust topology.

It was found that every robust topology could be constructed from the basic topological and functional modules shown in Figure 8.4, that is, every robust topology consisted of at least one E module, at least one B module, and at least one W module, although not every topology constructed out of at least one of each type of module was robust. The three module types, while topologically distinct, also correspond to three functional classes: the E modules give rise to bistability in E, the W modules give rise to bistability in W, and the B modules represent intercellular control between E and W that is required to delineate parasegment boundaries. Thus, the observed remarkable robustness of the segment polarity network is a direct consequence of its specific modular topology and modular functions.

8.2 Topological and Functional Aspects of Network Modularity

We have seen that the segment polarity network can be decomposed into modules, that is, smaller sub-networks that can be combined to produce the larger network by the addition of very few links or edges. In addition to this topological feature, the hallmark of the defined modules is that each module has a functional role that can be clearly delineated. In general, the existence of sub-networks that "look like" modules at the level of the connectivity pattern or topology, yet have well-defined

E modules   B modules   W modules

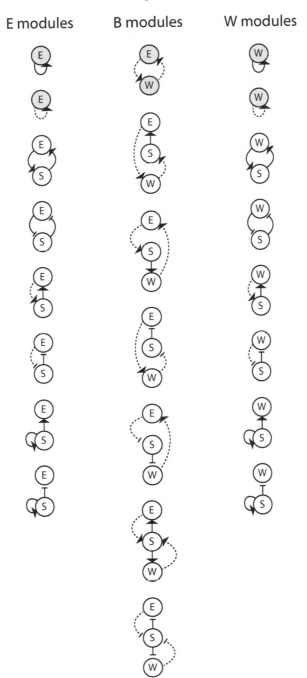

FIGURE 8.4: The three types of modules that constitute robust topologies for the segment polarity network (Ma et al., 2006). The E and W modules lead to bistability in E and W, respectively, while the B modules ensure mutual intercellular regulation of E by W, and vice versa. Arrows are depicted in a similar manner to those in Figure 8.3. The modules that are shaded gray accomplish their function without the S meta-node.

functional roles, is a nontrivial phenomenon. Even so, such modules are ubiquitous in biomolecular networks. It is therefore important to understand how to discover modules within networks and then to examine their biological functions.

8.2.1 Newman–Girvan Modularity

Given a large undirected network, such as a protein interaction network, how does one discover modules within it with no other information except its connectivity pattern (topology)? The answer, which may appear circular at first, is to partition the network into sub-networks or groups of nodes that maximize the *modularity* of the entire network. These sub-networks are the requisite modules. There are several definitions of modularity. In the specific case of protein interaction networks, we have already encountered the Markov clustering procedure in Chapter 5 to decipher protein complexes, which are a type of module. We now introduce a widely used and general definition of modularity due to Newman and Girvan (2004). This notion of modularity builds upon the intuitive idea that a network is modular if it contains sub-networks or groups (these groups are the putative modules) such that there are a large number of intra-group connections and a small number of inter-group connections; the larger the density of intra-group connections and the smaller the density of inter-group connections, the larger should be the modularity of the entire network.

Let us assume that we are given a partitioning of the undirected network into known groups or sub-networks, which are our putative modules. The *modularity* of this collection of modules is defined by Newman and Girvan in terms of the number of within-module edges minus the number of such edges expected in a random network with the same degree sequence. Suppose n_{kl} is the number of edges between module k and module l, and $n_k = n_{kk} + \sum_{l \neq k} n_{kl}/2$ is the total number of edges that connect to nodes in module k (the factor of $1/2$ arises because edges that connect nodes in module k with nodes in other modules are counted twice in the sum). Then $e_{kl} = n_{kl}/m$ is the fraction of the total number of edges in the network (which is here taken to be m) that connect module k to l, and $a_k = n_k/m$ is the fraction that connect to module k. Also, note that the sum of degrees of all nodes in the entire network is twice the total number of edges, that is, $\sum_i k_i = 2m$, because each edge is counted twice in the sum, and for similar reasons the sum of degrees of all nodes in module k is given

Box 8.2 Adjacency and Laplacian matrices.

The *adjacency matrix* **A** of an undirected network containing n nodes is an $n \times n$ matrix of 0s and 1s, in which $A_{ij} = 1$ if there is an edge connecting nodes i and j, and $A_{ij} = 0$ otherwise. The adjacency matrix thus provides a complete mathematical representation of the network. For directed networks, the adjacency matrix is an asymmetric matrix in which $A_{ij} = 1$ only if there is a directed edge from i to j.

A closely related matrix is the *Laplacian matrix* **L**, defined as **D** − **A**, where **D** is a diagonal $n \times n$ matrix with node degrees along the main diagonal, that is, $D_{ii} = k_i$. The Laplacian and adjacency matrices have a number of applications in the study of random walks on networks, which, in turn, have applications in the discovery of clusters and modules.

by $d_k = \sum_{i \in k} k_i = 2n_k$. Using these relationships, the Newman–Girvan modularity Q of the network can be expressed in a number of equivalent ways as follows:

$$Q = \sum_l (e_{ll} - a_l^2) \tag{8.3}$$

$$= \sum_l \left[\frac{n_{ll}}{m} - \left(\frac{n_l}{m} \right)^2 \right] \tag{8.4}$$

$$= \sum_l \left[\frac{n_{ll}}{m} - \left(\frac{d_l}{2m} \right)^2 \right] \tag{8.5}$$

$$= \frac{1}{2m} \sum_l \sum_{i,j \in l} \left[A_{ij} - \frac{k_i k_j}{2m} \right], \tag{8.6}$$

where $i, j \in l$ is shorthand notation which says that node i and node j belong to module l, and A_{ij} is the adjacency matrix of the network (Box 8.2). The reader should convince herself that the Newman–Girvan modularity of an unpartitioned network (or equivalently, a network in which all nodes belong to the same module) is zero.

The above definition of modularity assumes, of course, that the network has already been partitioned into modules. But how do we discover this partitioning in the first place? As stated above, the modules are themselves discovered by finding the partition that maximizes the modularity. While there are various algorithms to do this, one approach due to Newman (2006) that works efficiently and accurately even for large networks is to carry out the partitioning in a hierarchical manner: the network is first partitioned into two sub-networks; then each sub-network is further partitioned into two, and so on. This process is halted when subsequent partitioning of any sub-network does not increase the total modularity of the entire network. The algorithm to carry out the optimal partitioning at each stage is briefly outlined in Box 8.3.

There are various extensions and variants of Newman–Girvan modularity and the associated module-finding algorithm as well as other methods to find modules that have been applied to biological networks. The basic modularity measure can be modified for directed networks (Guimera et al., 2007) and networks with edge weights (Newman, 2004). A related measure that is particularly appropriate for networks with power-law degree distributions has been proposed by Fernandez (2007), and the unrelated procedure of Markov clustering that we previously encountered is simple, fast, and commonly used to analyze protein interaction networks. These, and many other methods to characterize modules, are exhaustively surveyed in the review by Fortunato (2010).

Now that we have discussed how to define and discover modules, let us explore some uses of module discovery. For example, can the Newman–Girvan procedure accurately discover modules corresponding to different biological functions? Protein interaction networks are natural candidates for the application of module-finding procedures. In an analysis of the efficacy of Newman–Girvan modules at discriminating between sets of proteins with different functions in yeast and human protein interaction networks, it was found that the discovered modules largely correspond to active protein complexes, consisting of groups of proteins that are known to cooperate to perform common functions (Dunn et al., 2005). The module detection method is insensitive to the addition and deletion of up to ~20% of random edges, thus confirming the robustness to false positive and false negative errors. The fact that modules can be detected robustly is a consequence of the inherent modular topology of protein interaction networks. Moreover, the fact that the detected modules correspond to protein groups of homogenous function is likely a consequence of protein interaction networks having evolved in such a way as to ensure that topological modules are also functional modules.

If protein interaction networks have evolved to be modular, should it

Box 8.3 Newman's module discovery algorithm.

Suppose we wish to initially partition an unpartitioned network in two partitions or modules in such a way so that the modularity of the partitioning is maximal. Based on Equation (8.6), we can define the so-called *modularity matrix* $\mathbf{B}$, as $B_{ij} = A_{ij} - k_i k_j / (2m)$. Given a partitioning of the network into two modules, we can also define an indicator vector $\mathbf{s}$, where $s_i = 1$ if node i belongs to module 1 and $s_i = -1$ if node i belongs to module 2. Then the modularity measure of Equation (8.6) can be expressed in matrix form as $Q = \mathbf{s}^T \mathbf{B} \mathbf{s} / (4m)$, where T denotes the transpose operation. The matrix $\mathbf{B}$ can be expressed in terms of the so-called *spectral expansion*, that is, in terms of its normalized eigenvectors $\mathbf{u_i}$, $i = 1, \ldots, n$, as $\mathbf{B} = \sum_{i=1}^{n} \beta_i \mathbf{u_i} \mathbf{u_i}^T$, where β_i are the eigenvalues of $\mathbf{B}$, arranged in decreasing order for convenience. Thus, the modularity Q can be written as

$$Q = \frac{1}{4m} \sum_{i=1}^{n} \beta_i \left(\mathbf{u_i}^T \cdot \mathbf{s} \right)^2. \tag{8.7}$$

The objective is to find the partitioning vector $\mathbf{s}$ for which the modularity Q is maximal. While this is a difficult problem to solve exactly, an approximate solution can be found by assuming that the sum in Equation (8.7) is dominated by the largest eigenvalue β_1. The eigenvector corresponding to the largest eigenvalue, $\mathbf{u_1}$, is called the *principal eigenvector* of the modularity matrix. An approximate solution to maximizing the modularity is then to find the vector $\mathbf{s}$ such that the dot product $\mathbf{u_1}^T \cdot \mathbf{s}$ is maximal. Because the entries of $\mathbf{s}$ are constrained to be ± 1, the dot product is maximized if one chooses $s_i = 1$ whenever the entry u_{1i} is positive and $s_i = -1$ whenever the entry u_{1i} is negative. This choice defines the optimal initial partitioning of the network into two partitions.

Further partitioning can be carried out in a similar manner by treating each of the initial partitions as independent networks, finding the principal eigenvectors of each modularity matrix, and choosing the sub-partitions based on the signs of the entries of these principal eigenvectors. The only caveat is that now the nodes in a partition g have intra- as well as inter-partition connections, and this feature leads to a modified modularity matrix $\mathbf{B}'$ whose principal eigenvector must be found. In terms of the original modularity matrix, the elements of this modified matrix are given by $B'_{ij} = B_{ij} - \delta_{ij} \sum_{l \in g} B_{il}$, where the additional term modifies the diagonal entries of the modularity matrix and involves a sum over all nodes within the partition g.

In this manner, the network can be recursively partitioned into modules. At each step, an existing partition is potentially subdivided into two partitions that are determined by the principal eigenvector of the appropriate $\mathbf{B}'$ matrix. The recursive procedure is halted when all entries of the principal eigenvector of a modified modularity matrix have the same sign. At that point, further partitioning cannot increase the modularity.

not be the case that ancient networks were less modular? Indeed, this appears to be true. Fernandez (2007) divided yeast proteins into five "age classes" based on their extent of sequence similarity with proteins in various kingdoms of life. In this division, yeast proteins that are homologous to proteins in the oldest life forms are classified as the oldest proteins, and so on. He then successively reconstructed more ancient networks by removing "young" nodes. For each reconstructed network, he computed its modularity. The modularity is found to steadily increase from a value of about 0.40 for the most ancient network (one in which all nodes have homologs in all ancestral groups) to about 0.55 for the present-day yeast network. The increase in modularity is accompanied by a progressive enrichment in intramodular hub proteins that preferentially do not connect to each other. Thus, although the most ancient protein interaction network has a power-law degree distribution, it possesses a high probability of connection between nodes of similar degree (and therefore between hubs) as compared to the present-day network, in which hubs tend to lie at the centers of modules and tend to connect to other nodes in the same module rather than to other hubs. The avoidance of hub-hub connections in the present network imparts to it a robustness to hub removal, as removal of a hub will disrupt the module that the hub is associated with rather than impact other hubs (and other modules). Because modularity is, in turn, connected to function, directed attack on a specific hub will impair the function associated with that hub rather than impair a host of functions. Again, we see the interplay between modularity and robustness at work.

Homogeneity of biological function within topologically discovered modules is not restricted to protein interaction networks alone. Metabolic networks also tend to be topologically and functionally modular, with the two sets of modules having high overlap. In a study of the metabolic networks of twelve different organisms, it was found that the overwhelming majority of modules discovered using the Newman–Girvan modularity measure contain metabolites mostly from one major pathway (Guimera and Amaral, 2005). Furthermore, in an analysis of intra- versus inter-module connections of each metabolite, it was found that non-hub metabolites whose few links were spread across different modules tend to be far more evolutinarily conserved than hub metabolites whose links are mostly confined within a single module. However, in general, hub metabolites are more conserved than non-hub metabolites. This is another example of evolutionary robustness:

non-hub metabolites that link to different modules are more important because disruption in their function would impair several cellular functions. By contrast, disruption of hub metabolites whose links lie within a single module would only lead to disruption of a single function. Evolution appears to have preferentially selected for organisms that carry functional forms of the former metabolite.

8.2.2 Functional Modules from Expression Data

Functionally relevant modules in networks are often discovered using other biological information in addition to connectivity data. A classic example consists of overlaying expression data on the yeast protein interaction network to reveal a model of organized modularity (Han et al., 2004). It was found that hub nodes in the yeast protein interaction network could be divided into two classes (Figure 8.5): those whose expression profile is strongly correlated with the expression profiles of their interacting partners (these hubs are termed *party hubs*), and those whose expression profile is only weakly correlated with the expression profiles of their interacting partners (these hubs are termed *date hubs*). Date hubs tend to have very different characteristics from those of party hubs: their interacting partners are expressed in a broad distribution of cellular conditions. The expression of interacting partners of party hubs are mostly confined to the same cellular conditions. The network rapidly fragments into smaller components when date hubs are systematically removed from it; party hub removal does not result in such a rapid fragmentation of the network. There are roughly twice as many genetic interactions involving date hubs as those that involve party hubs, suggesting that date hubs play a more central functional role than party hubs. When all date hubs are removed, the protein interaction network fragments into distinct sub-networks that can be interpreted as modules. These modules turn out to be functionally homogenous and correspond to components as diverse as protein complexes, molecular machines, and regulatory pathways (Box 8.4). Thus, the yeast protein interaction network may be said to follow a model of organized modularity in which functional modules are distinct topological units ruled by the party hubs, while date hubs link the modules to each other and therefore mediate the crosstalk between different functions.

Just as data integration has been an important theme in many topics discussed in this book, so it is with the discovery and analysis of func-

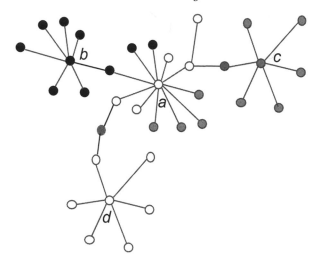

FIGURE 8.5: Party and date hubs. Protein nodes in the network shown above are shaded by their expression profiles: proteins having similar expression profiles are shaded in the same way. Of the four hub nodes (*a*, *b*, *c*, *d*), nodes *b*, *c* and *d* are party hubs because their expression profiles correlate strongly with those of their neighbors, while node *a* is a date hub because its expression profile is, on average, only weakly correlated with those of its neighbors.

tional modules. The above example of organized modularity involves integration of connectivity data and expression data to yield functional modules. In another example, expression data and knowledge of regulatory interactions was overlaid on protein interaction modules to find transcriptional factors that coregulate genes in two or more separate modules (Petti and Church, 2005), which is reminiscent of the data integration procedures used to predict protein-protein interactions in Chapter 3. The analysis of Petti and Church yielded a "super-network" in which nodes are modules and two modules are connected if they are co-regulated by at least one transcription factor. By combining protein interaction networks with expression and localization data, it is possible to discriminate between modules that correspond to protein complexes and functional modules in which all proteins participate in the same cellular process (Lu et al., 2006).

Box 8.4 Complexes, machines, and pathways.

By a *protein complex*, we generally mean an experimentally observed association of multiple proteins regardless of their biological functions. A *molecular machine* is a protein complex whose member proteins have experimentally observed biochemical functions that are highly orchestrated as a series of catalytic steps. For example, the eukaryotic RNA polymerase II is composed of approximately seventy-five proteins. The functions of these proteins include DNA sequence recognition, DNA strand unwinding, RNA synthesis, DNA strand rewinding, and protein movement along the DNA template, among many others. A second example of a molecular machine is the ribosome, which contains a number of different proteins as well as non-coding RNA molecules.

A *regulatory pathway* is a set of proteins that may or may not form a complex, yet the components together control the magnitude of a specific biological response, such as signal transduction or gene expression. The signal transduction pathways shown in Figure 1.2 are all regulatory pathways. One of them describes the following process: pheromone binding to a membrane receptor triggers a cascade of enzyme-catalyzed protein phosphorylation events ultimately leading to transport of an activated transcription factor protein into the nucleus where it causes global changes in gene expression. A second example that we encountered in Chapter 5 is the activation of the p53 transcription factor by a DNA damage-induced signal. As in the previous example, activated p53 enters the nucleus and causes global changes in gene expression.

8.3 Robustness Implications of Topology of Biological Networks

Robustness is the ability of a complex system to continue functioning despite significant perturbations to its components. At the network level, robustness can manifest itself in two different ways. Robustness can correspond to some phenotype of the system remaining approximately constant despite changes to system parameters or the addition of noise without changing its wiring diagram or connectivity pattern. By contrast, robustness can also correspond to the phenotype remaining constant despite structural changes in the wiring diagram that defines the system. We have encountered both types of robustness in the segment polarity network: the parasegmental expression pattern is robust to parameter changes in the differential equations underlying the dynamics of expression, and it is also robust to changes in the wiring diagram as long as the wiring diagram retains some fundamental functional modules. It is important to note that both kinds of changes—changes in parameters and changes in the wiring diagram—correspond to epigenetic[3] or genetic perturbations rather than physiological fluctuations.

With regard to robustness against perturbations of the wiring diagram itself, we have seen the implications of random or systematic perturbations of networks with different topologies. In Chapter 1, we discussed how power-law networks are robust to random removal of nodes but sensitive to directed attack on the hub nodes. In the previous section, we observed that the organized modularity model of protein interaction networks predicts that sensitivity of power-law networks to directed attack on hubs can be accounted for by the sensitivity to directed attack on the date hubs alone, while directed attack on party hubs does not significantly influence network topology. These studies probe the static robustness of a network, that is, they explore how the *structure* of the network responds to the perturbations such as node or edge removal.

The above conclusions make intuitive sense in light of how biological systems evolve. For example, physiological adaptation to changes in environmental conditions must necessarily entail a readjustment of network components to retain internal homeostasis (stability). External variables such as temperature would likely affect the structure and stability of numerous proteins in a manner analogous to a random perturbation process. Likewise, a power-law degree distribution of the protein-protein interaction network ought to make such a network more resistant to disintegration than if the interaction network were to have some other degree distribution. This is because there are few proteins with high degrees and numerous proteins with low degrees in a power-law network, which ensures that random perturbations such as node or edge removal would more likely affect the low-degree proteins, with relatively little impact on the overall function of the network.

The problem with a power-law topology, however, is its relative sen-

---

[3]Epigenetic perturbations are changes in gene expression levels or phenotype that are *not* caused by changes in the DNA sequence. An example of an epigenetic perturbation is DNA methylation, which is the addition of methyl groups to DNA. Highly methylated areas of DNA are generally transcribed to RNA at lower rates than nonmethylated regions, although there are some exceptions to this rule.

sitivity to other biological agents that can evolve to target the more vulnerable nodes, such as the date hubs. Certain bacterial viruses (e.g., the T7 bacteriophage) infect and rapidly reproduce within the host cell to make numerous copies of the virus, then reemerge by killing the host cell. Such viruses often inactivate certain host proteins (e.g., T7 inhibits the host-encoded RNA polymerase) that are protein-protein interaction network hubs (Nechaev and Severinov, 1998). Similarly, certain human viruses inactivate the p53 protein that is a hub in the human protein-protein interaction network (Portis et al., 2001). In fact, robustness analysis of the human protein-protein interaction sub-network centered around the p53 protein, whose function is adversely affected in over 75% of all cancers due to mutations in the gene that encodes it, indeed shows the sub-network to be robust to random edge removal but not to targeted edge removal (Dartnell et al., 2005).

Intracellular pathogens of higher plants elicit immune response by the host upon invading the host plant. The protein-protein interaction network of the immune system of the plant *Arabidopsis thaliana* (Mukhtar et al., 2011) consists of 926 protein nodes and 1,358 interactions. Plant pathogens encode some 20 to 30 proteins (called *effectors*) that manipulate the host cell immune system. Strikingly, the effector proteins of two different plant pathogens attack an overlapping set of proteins in the plant immune response network, and the targeted set of immune response proteins is significantly enriched for hubs (Figure 8.6). Thus, in a manner reminiscent of T7 bacteriophage inactivating the highly connected RNA polymerase of its host (see above) or phage lambda modifying the host RNA polymerase (discussed in Chapter 6), the plant pathogens also target "hub" proteins in the plant immune network. Here again we see an example of parallel evolution of the host and the pathogen, where the pathogens evolved to target the most vulnerable components of the network, thus compromising the robustness properties of the host.

8.4   Robustness and Network Dynamics

We have seen, here and in Chapter 6, that biological networks, such as the segment polarity network, the lambda gene expression network, and the protein interaction network of yeast cell cycle, are dynamic on physiological timescales because of time-varying expression levels of the

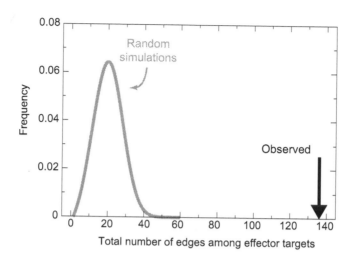

FIGURE 8.6: Pathogens target highly connected proteins in the immune network of *Arabidopsis thaliana*. The total number of direct edges that connect host proteins that are the targets of effector proteins encoded by pathogens are shown in this schematic figure. The black arrow on the right corresponds to the observed number of edges among the targeted host proteins in *Arabidopsis*, while the curve on the left is the distribution of this number assuming random connections among the host proteins. This distribution was generated empirically on the computer by randomly shuffling the edges between nodes, keeping the degree distribution of host proteins intact and identical to the experimentally determined one. A total of 15,000 different random networks were generated by the shuffling process, each of which had a different total number of edges among the effector targets. The figure shows that the observed number of edges among effector targets is far higher than that expected by chance. Adapted with permission from Mukhtar et al. (2011).

mRNA and proteins. To what extent can the robustness of biological networks be attributed to the dynamical properties of the networks themselves?

Unlike the computer simulations performed to address robustness of the segment polarity network, the question of dynamical robustness to genetic perturbations has been addressed experimentally in the case of bacteriophage lambda, where it was found that the switch from the lytic to the lysogenic cycle (see Chapter 6) is highly robust to mutations in the operator region OR that is occupied by the two regulatory proteins cI and Cro (Little and Shepley, 2006), a property that is crucial to the lysis-lysogeny switch. Despite significant mutational changes that are predicted to alter the binding efficiencies of cI and Cro proteins by one to three orders of magnitude, the significance of these changes to the viral biology was minimally disruptive, varying at most by a factor of three. A conclusion from this study was that the lambda gene regulatory network is not delicately poised at the edge of chaotic behavior but is highly optimized and robust to random mutations in its important components.

The source of the observed genetic robustness of the phage lambda regulatory circuit was later examined by constructing a mathematical model. This model incorporated stochastic white noise components into a system of coupled differential equations that described the concentrations of cI and Cro proteins in the bound and unbound states on the operators, and their mutual interactions. It was concluded that a crucially important factor for the observed robustness is the positive feedback of cI protein on transcription of the cI gene, which is enhanced by the cooperative binding of the cI proteins to the operator sequences (Zhu et al., 2004). This is very similar to the robustness encountered in the segment polarity network, where positive feedback on the *engrailed* and *wingless* genes plays a crucial role.

The above discussion points to the importance of examining the general dynamical properties of biological interaction networks as the source of genetic robustness. While the existence of positive feedback or auto-activation loops is crucial for the robustness of small metabolic and regulatory networks, this observation does not clarify the topological features that contribute to the robustness of larger networks. We now describe how the power-law nature of the degree distribution of large biological networks contributes to the dynamical robustness of network properties.

As a simple approximation, imagine a large network with N nodes, each of which can exist in one of two states: the ON or "Activated" state, and the OFF or "Inactivated" state. This is a dynamical Boolean network of the type we have already encountered. The state of a node at a certain time point is determined in a stochastic manner by the state at the previous time point of all nodes connecting to it. Denoting the state of node i at time t by $\sigma_i(t)$, the dynamics can be expressed as

$$\sigma_i(t+1) = f_i\left(\sigma_{i_1}(t), \sigma_{i_2}(t), \ldots, \sigma_{i_{k_i}}(t)\right), \qquad (8.8)$$

where f_i is a different random Boolean function of its inputs for every node i, and k_i is the degree of node i. Thus, there are 2^{k_i} possible input states for the function f_i and the function outputs a value of 0 (OFF) or 1 (ON) for each state, with the choice determined in a random manner. However, once the choice of a 0 or 1 output value for each input state is made, it remains constant for all time.

The state dynamics of a Boolean network eventually settle to cyclical behavior, where a cycle consists of a set of states that is visited periodically. In the biological context, a cycle corresponds to a set of gene expression patterns corresponding to intracellular tasks. Typically, these cyclical dynamics have two regimes: a chaotic regime in which even the perturbation of the state of a single node can make the system jump from one cycle to another, and a robust regime in which perturbations die out over time. The robust regime is therefore insensitive to initial conditions.

What parameters dictate whether the network is in a chaotic or robust regime? The relevant parameters that characterize a Boolean network are a parameter ρ that corresponds to the overall probability that the Boolean functions f_i output a value of 1 (thus, $1 - \rho$ is the probability that they output a value of 0), and a second parameter that characterizes the topology or wiring diagram of the network. If the network is a random network, a meaningful choice of this second parameter is the average node degree K. On the other hand, if the network is a power-law network with exponent γ (in which the variance of the degree is infinite for $\gamma \leq 3$), the more meaningful choice of the topological parameter is the power-law exponent γ. Aldana and Cluzel (2003) found that while most choices of ρ and K for random graphs lead to chaotic dynamics (in particular, for a mean connectivity value of $K = 20$—which is representative of the human regulatory network— robust behavior is attained only for ~5% of values of ρ), a large range of choice of ρ and γ lead to robust behavior for power-law networks. In particular, every choice of ρ for $\gamma > 2.5$ in power-law networks leads to robust dynamics. However, it is also true that every choice of ρ for $\gamma < 2.0$ leads to chaotic behavior in power-law networks. Thus values of γ between 2.0 and 2.5 correspond to a critical regime in which robustness is dependent on the precise value of ρ. Interestingly, it turns out that the power-law exponents of most real networks (not just biological ones) lie between 2.0 and 2.5. In the robust regime, it turns out that robustness to perturbations is strongly dependent on the node that is perturbed. Perturbations of high-degree nodes spread through the network and in fact lead to chaotic behavior, while perturbations of the states of low-degree nodes die out. Because a power-law network possesses many more low-degree nodes than high-degree nodes, this phenomenon is consistent with the overall robustness and is also consistent with the finding that power-law networks are topologically robust to random node removal while being sensitive to directed removal of high-degree nodes. If we assume that Boolean dynamics approximate the true dynamics of expression to some degree, these results show that power-law networks are not just topologically robust to random attack, but also *dynamically* robust.

8.5 Network Robustness by Rewiring

We have so far discussed mechanisms of evolutionary adaptation of interaction networks to selection pressure, and have seen that both network topology and network dynamics are important for the evolution of robust features. It appears that selection pressure on the functionality of network components constrains the network topology feature-space to a relatively narrow feature, approximating a power-law degree distribution. This however leaves an important question unanswered: Is it possible that embedded within an extant interaction network are alternate pathways that appear under severe stress to the network? The motivation for raising this question comes from an anthropomorphic analogy to contingency plans in the face of unforeseen perturbations, which are often embedded in human social networks. Take, for instance, the backup plan for communication networks in the event of large natural disasters. Such backup networks are generally not in continuous use, but their very existence is designed to impart robustness. Are there equivalent backup networks of interaction in the cells that come into operation at times of severe threats to survival?

To address this question, one needs to first work out an interaction network that is operative under normal conditions of growth, then perturb or stress the organism and work out the interaction network under the perturbed condition. Indeed, this was done for the network of genetic interactions in yeast when grown under normal conditions and under conditions of acute stress produced by adding a chemical that causes damage to the DNA (Bandyopadhyay et al., 2010). Recall

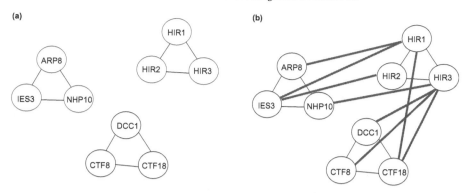

FIGURE 8.7: Rewiring of a genetic interaction network under stress. (a) In the absence of extensive DNA damage, there was no evidence of genetic interactions among the nine genes represented as circles. However, they form three modules of protein-protein interactions (thin lines). (b) In the presence of extensive damage to the DNA (as occurs, for example, when yeast cells are grown in the presence of a chemical that causes DNA breaks in yeast), a number of genetic interactions appeared (thick lines). Adapted and simplified from Bandyopadhyay et al. (2010).

that a genetic interaction network can be constructed out of different types of gene-gene interactions. The specific interaction type used in this study was that of epistasis interaction, in which when a pair of genes is simultaneously deleted, their combined effect is either more (positive epistasis) or less (negative epistasis) than the combination of their individual effects (see Chapter 2).

Strikingly, it was found that many of the epistatic pairs changed when they were tested in the presence of the chemical. However, many novel epistatic pairs emerged in the presence of the chemical. It appears that the genetic interaction network is rewired in the presence of the chemical that is expected to cause severe damage to the DNA. An alternate, albeit trivial, interpretation of this result is that the chemical caused numerous mutations in the background, some of them suppressors of synthetic lethal pairs and others causing novel lethal mutations. This interpretation was not credible given the high frequency at which such specific interactions (or lack thereof) were exhibited. In fact, many of the novel interactions made sense in terms of DNA repair pathways required for handing damages to the DNA by the chemical (Figure 8.7). There is a difficulty with the above view, however. Because it is widely thought that evolution selects the maintenance of genes having a spe-

cific adaptive value, genes that participate in "contingency plans"—
that is, those that have alternate functions—would appear to have no
selective advantage for such alternate functional specialization. One
can argue against this by assuming that the perturbing factors, in this
case damage to the DNA, are fairly frequent and occur naturally over
the course of evolution, causing frequent lethal impact during normal
cellular growth processes, such that the contingency plans are regularly
utilized and therefore contribute to survival frequently in the population. Yet another argument for the existence of "contingency" genes is
that the alternate functions actually arise out of the emergent behavior
of collections of genes, utilizing remnants of ancestral functions that are
not yet lost through mutational erosion.

Chapter 9

Networks and Disease

One of the most important advances in understanding the molecular basis of diseases and the promise of novel therapeutic discoveries has been the use of network data as well as network paradigms. As we show in this chapter, these approaches are leading to the identification of important genes that are implicated in various diseases, the understanding of common origins of disparate diseases, disease comorbidity, prediction of patient response to disease states, and the design of novel treatment methods.

Before we embark on an exploration of the connections between network biology and disease, we must first understand the concept of a *disease gene*. A disease gene is a gene in which specific mutations, or whose abnormal expression, cause the disease in question. A disease may have one or more disease genes associated with it. The abnormal protein sometimes coded by the disease gene is often termed a disease-causing protein. There is growing realization that susceptibility to nearly all human diseases, including infectious diseases, may have at least some genetic component. A comprehensive atlas of genetic diseases and the genes associated with them is provided by the Online Mendelian Inheritance in Man (OMIM) database (http://www.omim.org). As of August 2012, the OMIM database contained over 21,000 entries. In some instances, different sets of mutations in the same gene may give rise to different disorders, just as most disorders are polygenic (i.e., caused by mutations in more than one gene). This "multi-gene-multi-disease" association immediately suggests a network-like paradigm for understanding the relationships among human diseases. Before we delve deeper into this paradigm, however, it is instructive to further understand how a disease can be caused by single gene mutations. Such diseases are termed *Mendelian diseases*. Examples are the so-called in-born errors of metabolism, such as phenylketonuria (PKU). In this disease, the affected patients cannot metabolize a particular amino acid effectively, leading to toxic accumulation of a degradation product and consequent disorders. Other examples are sickle-cell anemia, which is caused by a mutation that produces an aberrant hemoglobin molecule (that is unable to carry sufficient oxygen) in the red blood cells, and hemophilia, which causes a blood clotting disorder.

Genetic diseases can be due to a dominant or a recessive mutation, depending on whether one or both copies, respectively, of the disease gene (one inherited from the mother and one inherited from the father) are required for the individual to be afflicted with the disease. Mendelian diseases are further classified as *autosomal* diseases if their disease gene resides on one of the twenty-two non-sex chromosomes, *X-linked* if their disease gene resides on the *X*-chromosome, and *Y-linked* if their disease gene resides on the *Y*-chromosome. Thus, *autosomal recessive* diseases are those in which both copies of the disease gene (that resides on the two homologous non-sex chromosomes) must be mutated for the individual to be affected. In such diseases, it is possible that two parents, each possessing one copy of the disease gene (and therefore, not afflicted by the disease) can have a child who has two copies of the disease gene and is therefore afflicted. Examples of autosomal recessive diseases include cystic fibrosis and sickle-cell anemia. In *autosomal dominant* diseases, on the other hand, one copy of the mutated gene is sufficient for the disease to be expressed. It is therefore not possible for an individual to be afflicted with an autosomal dominant disease unless at least one of his/her parents is afflicted with it[1]. Huntington's disease is an autosomal dominant disease.

X-linked recessive diseases are only expressed in females if they carry two copies of the disease gene, one on each chromosome. Only one copy of the disease gene is sufficient for the disease to be expressed in males, because males carry only one *X* chromosome. Thus, *X*-linked recessive diseases occur more frequently in males than in females. Hemophilia and color blindness are examples of *X*-linked recessive diseases. *Y*-linked diseases occur exclusively in males and are passed from father to son. Because there is only one *Y* chromosome (which exists only in males), the notions of recessive and dominant forms do not usu-

---

[1]It should be noted, however, that some autosomal dominant diseases have low *penetrance*, which means that not all individuals who carry the mutated form of the disease gene are afflicted by the disease. In rare cases, an autosomal dominant mutation can arise in the germline or within the first few cell divisions of the embryo. In such cases, the parents of the afflicted child may not carry the disease-causing mutation.

ally apply to Y-linked diseases. The most well known Y-linked diseases are those associated with mutations of the SRY (Sex-determining Region Y) gene. This gene codes for a protein that initiates male sex determination. Thus, mutations in this gene in men lead to an externally female-like appearance with under-developed gonads, a condition known as *Swyer syndrome*.

9.1 Disease Loci Identification and Mapping

As the above examples show, Mendelian diseases are relatively simple to identify because they are inherited in precisely predicted patterns in the human lineage, or *pedigree*. To know that a gene is responsible for a Mendelian disease is therefore straightforward in principle, but to identify the causal gene mutation is another matter. The identification of the sickle-cell hemoglobin gene as the carrier of the causal gene mutation in sickle-cell anemia required a large body of work in human physiology relating the anemic condition and the peculiar shape of the red blood cells (the classic "sickle" shape) in the diseased individual to the variant hemoglobin protein. Final confirmation of the causal gene mutation required sequencing the hemoglobin proteins purified from normal and sickle-cell patients, and demonstration of the single amino acid changes in the variant hemoglobin proteins in patients with specific mutations. Such successes were rarely possible for Mendelian disease genes until progress in molecular biology techniques such as the ability to amplify segments of DNA from the genome by polymerase chain reaction (PCR) and subsequent sequencing of these amplified DNA molecules was carried out. Nowadays it is possible to sequence the entire genome of family members with identical and nonidentical twins in which a Mendelian disease has occurred and thus find the single gene mutation that is responsible for the disease. One such disease gene that was identified by this method by a team of scientists led by D. Galas (Roach et al., 2010) causes Miller's syndrome, which is characterized by abnormal facial and limb features. As of this writing, there are over 2,600 genes currently known in which mutations are known to cause a disease.

For many diseases, however, it is difficult to track down the responsible DNA sequence variants. A common reason for this is that multiple gene variants make incremental contributions to a disease. Moreover, there may be variants in other genes that could potentially mask or suppress the deleterious effects of a disease-causing variant gene in a manner analogous to that of suppressor mutations or epistasis effects discussed in Chapter 2. In these cases, it is important to track down the regions (loci) of the DNA that are closely linked to the inheritance pattern of the diseased phenotype.

Even in the case of Mendelian diseases, it is sometimes hard to narrow the search for the disease gene to a small group of genes because of lack of information on altered protein function or biochemical or physiological disease mechanisms. In such cases, knowledge of the disease locus on the chromosome without specific knowledge of the disease gene can still provide crucial information.

In a classic example, two methods were used in parallel to track down a key risk factor gene for human colorectal cancer. To begin with, family pedigree studies indicated that there might be a gene variant running in certain families that increased the likelihood for a specific type of colorectal cancer, but it was not a clear-cut case of Mendelian inheritance. It was found that in susceptible patients with tumors, many of which developed into cancers, the genome was somewhat unstable: there were high-frequency mutations of DNA sequences with short repeated sequence stretches—this is the so-called *micro-satellite instability* in the genome. Tumors with such instability appeared to arise among individuals with a tendency to inherit a short stretch of a particular chromosomal DNA segment. This was possible to figure out because human chromosomes contain numerous differences in DNA sequence between the two sets of homologous chromosomes inherited from the mother and the father. These sequence differences are called *markers*, because they define "markings" on the chromosomes that can be mapped with respect to one another. In the respective parents germline cells (i.e., precursor cells of the egg or the sperm), the homologous chromosomes exchange parts by the process of recombination. Thus, the closer a pair of markers is, the less frequently the region of DNA between the marker pair is exchanged.

If a disease gene is located between a given pair of markers, then a parent who carries the disease-causing mutation on one chromosome would harbor some egg or sperm cells in which the marker pairs flanking the disease gene have exchanged (or recombined) and other cells in which they have not, at a frequency depending on the separation between the markers. It can be shown (Box 9.1 and Figure 9.1) that if the

Box 9.1 Dependence of recombination frequency on distance between genetic markers.

Suppose a is a disease-causing mutation (its wild-type allele is denoted by "+" in Figure 9.1), and b is a marker allele (with wild-type version B) located a distance d from a. Assuming that crossover exchanges are randomly and uniformly distributed across the chromosomes, the larger the separation d, the higher the probability of a crossover exchange occurring between the two loci a and b during meiotic cell division. Assuming further that crossovers are independent events (i.e., one exchange does not interfere with another) and that the mean number of exchanges between the two loci are small, then the number of exchanges between loci that are a distance d apart can be modeled by a Poisson process with mean d. That is, the probability of n crossover exchanges occurring within the distance d is $d^n \exp(-d)/n!$. Because recombination can only occur if the number of exchanges is odd, the probability or frequency of recombination is given by

$$r = \sum_{n \text{ odd}} \frac{d^n \exp(-d)}{n!} = d \exp(-d) + \frac{d^3 \exp(-d)}{3!} + \frac{d^5 \exp(-d)}{5!} + \cdots$$

$$= \exp(-d) \sinh(d) = \frac{1}{2}\left(1 - \exp(-2d)\right). \tag{9.2}$$

frequency of recombination is small (typically less than 20% per cell) and the exchange points are randomly distributed, then the distance of separation d between the pair of markers is related to recombination frequency r, as

$$r = \frac{1}{2}\left(1 - e^{-2d}\right). \tag{9.1}$$

Therefore, the closer a disease gene is to a marker, the greater is the chance that the disease causing mutation will be co-inherited with one specific version of the marker. By observing the frequency of inheritance of many such markers, in which the distance between two markers can in principle be made arbitrarily small (up to the limit of the two markers being adjacent DNA bases), one can map all such intervals between the markers. Therefore, the disease-causing mutant varieties can be located close to one or more such markers. This way of genetic mapping was employed over a pedigree of individuals in a family to map the disease-causing mutation for colorectal cancer to a region on human chromosome 2.

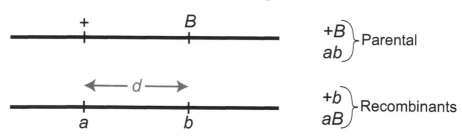

FIGURE 9.1: Marker separation and recombination. The chromosome shown on top contains wild-type alleles + and *B* while the bottom one contains their mutants *a* and *b* prior to recombination. After recombination, the strands are partially exchanged, leading to +*b* on the top chromosome and *aB* on the bottom chromosome, as indicated on the right. The intra-chromosomal distance between the markers is denoted by *d*.

From another direction, research on mechanisms by which DNA base pair mismatches are corrected had pointed out a crucial gene in *Escherichia coli* called *mutS* that is required for the repair of DNA base-pair mismatches that arise due to errors in DNA replication. Researchers led by R. Kolodner (Fishel et al., 1993) also identified homologs of *mutS* in yeast, and ultimately in humans. In yeast, mutations in two genes that are *mutS* homologs, called MutSHomolog1 (*MSH1*) and likewise *MSH2*, were discovered to cause micro-satellite instability in yeast, reflecting a similar mechanism of DNA sequence instability in both yeast and human colorectal tumors. Very soon it was discovered that the human version of *MSH1*, or *hMSH1*, mapped to the same region of chromosome 2 where the susceptibility gene to colorectal cancers was originally mapped. It remained for the Kolodner group to confirm by amplifying a bit of tumor DNA and sequencing the *hMSH1* gene residing there that there was indeed a mutation in the *hMSH1* gene in such cells. This was the first-ever cancer susceptibility gene to be identified.

The above account contains two important messages. The first is that human disease genes can be mapped with respect to DNA sequence landmarks on chromosomes. This is known as mapping of the disease gene locus. The second is that it is important to know something about the function of the gene to enable the identification of a disease gene because numerous genes may exist within the mapped interval of human chromosomes, many perhaps with DNA sequence variants that may or may not be causal to the disease. This is precisely one of the advantages of establishing an interaction network of human genes or proteins. If a human gene or protein with an unknown function is shown to interact with a set of other genes or proteins, then the function of this gene can be inferred from the functions of its interacting partners ("guilt by association").

9.2   Network as a Paradigm for Linking Diseases and Disease Genes

A global view of human genetic diseases and the genes that cause them can be constructed in network form using the data in OMIM (Goh et al., 2007) to reveal interesting disease-gene associations. In its basic form, this network is a *bipartite* network, which is a network with two types of nodes such that nodes of one type only connect to nodes of the other type. Here, the two types of nodes are disease nodes and gene nodes. A gene node is connected to a disease node if the gene in question is known to be a disease gene for the disease in question. This network consisted of 1,777 disease genes and 1,284 diseases. It is possible to project this bipartite network to obtain two "unipartite" networks: a *human disease network* (HDN) in which nodes are diseases and two diseases are linked if they share at least one disease gene, and a *disease gene network* (DGN) in which nodes are genes and two genes are linked if they are disease genes for the same disease. It turns out that both of these networks display very interesting clustering properties.

Independently of the network construction procedure, the diseases in OMIM were separately grouped into twenty-two separate disorder classes based on the physiological system affected by the disease. It turns out, perhaps not suprisingly, that both the DGN and HDN naturally possess clusters that correspond approximately to disorder classes, although there are a number of inter-class links that might suggest hypotheses about the related origins of disparate diseases or about relationships between disease genes. This local functional clustering leads to a global network topology in which the largest connected component is smaller than what one would expect for a random network. Furthermore, different disorder classes cluster in different ways. One of the most densely connected clusters in the HDN turns out to be the cancer cluster (Figure 9.2) due to the many genes associated with multiple cancer types. On the other hand, diseases associated with the breakdown of metabolism do not form a distinct cluster in the HDN because of their low genetic heterogeneity. Nonobvious connections between genes in the DGN could uncover novel similarities between genes that have no apparent similarity at the sequence, protein structure, or known function levels. Similar conclusions hold for diseases; for example, obesity is linked in the HDN to seven other diseases, including the well-known link to diabetes, but also to asthma, lipodystrophy, and glioblastoma (Barabási, 2007).

Goh et al. (2007) further examined the issue of robustness of the HDN and the DGN by adding data from the OMIM where disease phenotype is mapped more weakly to specific genes. In this broader dataset that now has 2,765 genes, while certain genes are known to be implicated in certain diseases, specific mutations that are causative of disease are not known in general. It turns out that the overall clustering and connectivity properties of the larger networks do not change upon enlargement, indicating that the qualitative conclusions about the structure of the HDN and DGN are robust.

A particularly important sub-network of the HDN is the so-called mitochondrial disease network. In the context of the discussion of the cellular basis of life in Chapter 1, we touched upon the evolution of mitochondria by mutualism between two or three different kinds of organisms, and made the point that cellular components exhibit intricate levels of interaction. When some of these interactions go awry, malfunction of these parts causes diseases with complex manifestations. In humans, a large number of diseases occur due to malfunction of the mitochondria. In Figure 9.3, a section of the mitochondrial disease-gene association network is shown, in which a number of human diseases that are caused by genes that control mitochondrial function, some of them even encoded by mitochondrial genes, are linked to one another via these causative genes. In this network, a gene node is connected to a disease node only if it is known to cause the disease and the gene is known to be related to mitochondrial function (or encoded by the mitochondrial genome). This particular network reveals a cluster of disorders sharing causative genes, all of which encode proteins that are known to help other proteins to fold under physiological conditions. MELAS syndrome, Leigh syndrome, and MERRF syndrome particularly appear to share more of the causative genes than other diseases. It would be interesting to analyze the types of genetic interactions shared by these causative genes, and how these interactions become defec-

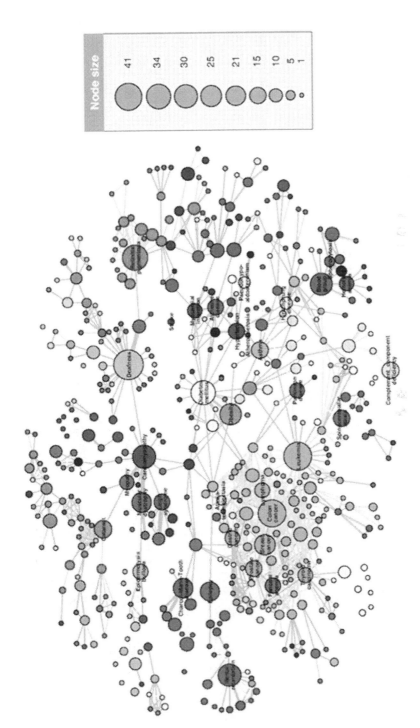

FIGURE 9.2: A portion of the Human Disease Network. The cancer cluster is at the lower left corner, in light gray. Node sizes are based on the number of genes implicated in the disease, as shown on the right. Reproduced with permission from Goh et al. (2007). [©2007 National Academy of Sciences, U.S.A.]

tive in individuals that carry mutations in these genes. A majority of mitochondrial disorders affect individuals at a very young age, even in infancy, and are vastly debilitating. There is also a high probability that the causative genes might be essential genes considering the severity of the clinical presentation of the diseases they cause. These questions are largely open at the time of writing.

Thinking about human diseases using the network paradigm is even more valuable when one overlays disease networks with the molecular networks that we have discussed in this book, such as protein-protein and metabolic networks. Protein-protein interaction networks, for example, help broaden the concept of a disease gene and may reveal associated genes that have important roles to play in disease. This phenomenon is discussed in greater detail in the following section. Furthermore, protein interaction networks can be viewed as a substrate for analyzing relationships between disease genes and other types of genes. By overlaying disease genes and aging genes (these are genes known to regulate the aging process) on protein interaction networks, Wang et al. (2009) found that the set of disease genes has greater overlap with the set of aging genes than expected by chance, and that disease genes are significantly closer to aging genes in protein interaction networks than expected by chance. This type of network-based analysis helps pave the way for the generation of novel hypotheses concerning the relationship between phenotypes such as aging and disease.

Perhaps surprisingly, new insights may be available by the analysis of disease networks and molecular networks in conjunction with *social* networks. As pointed out by Barabási (2007), obesity, for example, has a social component in addition to a genetic one. It is more likely for a person to be obese if his friends are, a phenomenon that causes obesity to be clustered among social communities. A simultaneous understanding of the interplay between molecular, disease, and social networks is likely to lead to important advances in medicine. While we do not discuss social networks in this book, this chapter elucidates ways in which diseases and disease networks can be overlayed with other molecular networks to yield fresh insights into the nature of disease.

It turns out that most network-based analyses of diseases employ protein interaction networks. As we discussed in Chapters 1 and 7, protein interaction networks, at least in baker's yeast, have properties that relate to phenotype in a simple way: hub proteins in protein interaction networks are enriched for proteins that are essential for life,

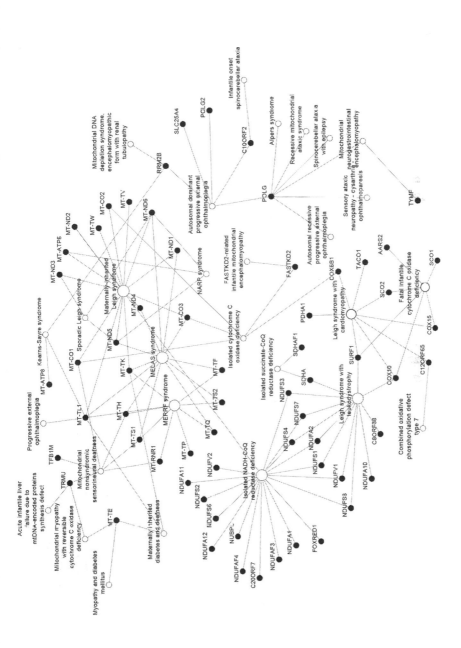

FIGURE 9.3: A portion of the mitochondrial gene-disease association network. The largest connected component of this bipartite network is shown. Unfilled nodes represent mitochondrial disorders, while filled nodes represent their causative genes. The causative genes are specifically associated with protein folding. Courtesy of S. Talele.

and it is possible to predict synthetic genetic interactions from protein interaction data with reasonable accuracy. These considerations can be extended to address the following questions that are based on human protein interaction networks. Can topological properties of a protein in a protein interaction network be used to characterize the corresponding gene as a disease gene? What more can we learn about diseases and disease genes by analyzing the network properties of the corresponding proteins? These are the questions we address in the following two sections.

9.3   Network-Based Prediction of Disease Genes

The idea of using human protein interaction data to predict disease genes stems from the observation in baker's yeast that proteins that are members of the same complex tend to produce similar phenotypes upon knockout of their corresponding genes. While there are a number of disease gene prediction methods that use protein interaction data, all of them, to degrees of varying sophistication, build upon this basic observation. Here, we describe some of the representative prediction methods. These fall into two broad categories: methods that use the network properties of known disease genes to predict new ones, and *ab initio* methods that combine data from differential expression studies with network properties to predict new disease genes.

9.3.1   Predicting Disease Genes from Known Disease Genes

Oti et al. (2006) used known disease genes culled from the OMIM database in combination with protein interaction data to predict unknown disease genes. To do so, they culled disease data from the OMIM for diseases that have at least one known disease gene and at least one disease locus lacking an identified disease gene. The disease loci studied by Oti et al. contained a median of 88 genes per locus, and 383 diseases satisfied their criteria at the time the study was carried out, with 1,195 disease loci having identified disease genes and 432 loci lacking identified disease genes. Next, they generated a human protein interaction network (HPIN) by combining individual protein-protein interactions culled from the literature, high-throughput experiments, and inferred protein-protein interactions by homology to the corresponding proteins in other eukaryotes, namely, the fruit fly, baker's yeast, and the worm.

This combined data were used to predict new disease genes for a specific disease in the following manner:

- For each known disease gene, protein interaction partners in the HPIN of the corresponding disease protein are found.

- The chromosomal locations of the genes coding for the interaction partners are determined using gene location data.

- These chromosomal locations are checked to see if they fall within a disease locus of the disease in question.

- Genes corresponding to interacting proteins that fall within a disease locus that lacks other disease genes (for the same disease) are predicted to be disease genes.

The success of this relatively simple method to predict disease genes depends crucially on the quality of the underlying protein interaction data. Among the various protein interaction datasets that Oti et al. (2006) used, the most reliable one was the set consisting of individual human protein-protein interactions manually curated from the literature. In this dataset alone, it turns out that nearly 60% of interactors of a disease gene correspond to genes that fall in a disease locus of the disease in question. This number falls to 9% when protein-protein interactions inferred from high-throughput interactions in the worm alone are used. Interestingly, the success rate is higher when using interactors inferred from yeast high-throughput experiments than from human high-throughput experiments. The reason for this is not completely understood, but may be related to high false negative rates in human interaction two-hybrid data.

The results of Oti et al. (2006) led to a landmark study by Fraser and Plotkin (2007), who generalized the notion of predicting disease genes to predicting genes that, upon knockout or mutation, shared a similar phenotype to other genes whose knockout/mutation phenotype was known. Their study was therefore about the prediction of *phenotype pairs*, that is, pairs of genes that share a similar phenotype upon mutation/knockout. In this study, initially carried out in yeast because of the availability of a large amount of genome-scale, high-throughput data, they examined the question of finding the best predictor of phenotype pairs among a large set of potential predictors that included genetic interactions, shared protein domains, shared transcription factors, co-expression of mRNA under different conditions, similarity of phylogenetic profiles, and several datasets of protein-protein interactions and protein complexes. Remarkably, they found that, when each predictor was considered separately for its predictive value, the best predictor of phenotype pairs was participation of the two proteins in the same protein complex. The other top predictors all corresponded to protein-protein interactions between members of the pair. Furthermore, it turns out that the predictive accuracy of every predictor other than shared protein complexes increases when the predictor data are intersected with co-expression data. Because we know that intersection with co-expression decreases the false positive rates in protein interaction data, these results clearly show that phenotypical similarity is best predicted by accurate protein interaction data, as exemplified by protein complexes, and the prediction accuracy increases as the quality of the protein interaction data increases.

Fraser and Plotkin (2007) used the lessons they learned from yeast data analysis to predict disease genes in humans by intersecting co-expression data with error-prone protein interaction data. Using manually curated protein interaction data culled from the literature, for example, they found twice as much enrichment for disease gene pairs among physically interacting proteins when these interactions were filtered by co-expression data (with a co-expression correlation cutoff of 0.3) as compared to the case where no such filtering was done.

Insights from the above-mentioned studies were further extended by Wu et al. (2008) and used to construct a prediction framework for disease genes that is based on the notion that phenotypic similarity between diseases is strongly correlated with closeness between the causative genes on a protein interaction network. Thus, given a disease and a gene, the method involves the computation of a *disease phenotype similarity profile* and a *gene closeness profile*. If these two profiles are strongly correlated, the gene is predicted to be a disease gene for the disease in question.

How are the two profiles defined? We briefly outline the steps here. Given a disease phenotype p and a gene g, the disease phenotype similarity profile S_p is a numerical vector of similarities between p and all other disease phenotypes:

$$S_p = (S_{pp_1}, S_{pp_2}, \ldots, S_{pp_n}), \tag{9.3}$$

where the similarities $S_{pp'}$ between two phenotypes p and p' are computed by appropriately assigning numerical values to phenotypes [while the procedure of assigning numbers to phenotypes is somewhat cumbersome and beyond the scope of this book, see, for example, van Driel et al. (2006)]. In an analogous manner, the gene closeness profile C_g is a numerical vector that quantifies closeness between g and all disease phenotypes using path lengths on a protein interaction network. Suppose the set of all known disease genes for the phenotype p is denoted by $G(p)$, and suppose that $L_{gg'}$ denotes the distance (e.g., shortest path length) in a protein interaction network between the proteins that genes g and g' code for. Then the gene closeness profile is defined as

$$C_g = (C_{gp_1}, C_{gp_2}, \ldots, C_{gp_n}), \tag{9.4}$$

where the closeness between a gene and a phenotype is further expressed in terms of the closeness between the gene and all other genes responsible for the phenotype on a protein interaction network:

$$C_{gp} = \sum_{g' \in G(p)} e^{-L_{gg'}}. \tag{9.5}$$

The correlation between the vectors S_p and C_g is predicted to be a measure of the propensity for gene g to be a disease gene responsible for the phenotype p. Given a phenotype p, it is possible to evaluate this prediction method by computing the requisite correlations between S_p and the closeness profiles of all known disease genes and subsequently ranking genes on the basis of the strength of their correlation with S_p. The method can be judged as successful if the correct disease gene appears as the top-ranked gene in this procedure. While the detailed performance statistics depend upon how one defines network distance and on the correlation cutoff that stipulates a rank 1 prediction to be significant, this is overall a promising method to predict new disease genes. Moreover, it nicely illustrates and exploits the correspondence between phenotypic closeness and closeness on a protein interaction network.

The notion of protein interaction network closeness can be defined in a more global manner rather than by considering only direct interactions and shortest paths between a candidate gene and a known disease gene in a protein interaction network. For example, Köhler et al. (2008) constructed a random walk through the human protein interaction network that was constrained to stay in the vicinity of disease genes with high probability. The most frequently visited nodes (proteins) in this random walk were then predicted to correspond to new disease genes.

Specifically, the random walk was constructed to begin with equal probability at any of the known disease-causing proteins for a particular disease. Subsequently, the walk could either jump to a protein that directly interacts with the starting protein (with probability $1 - r$) or to any of the known disease-causing proteins (with probability r). This process was repeated until convergence. Given the column-normalized adjacency matrix $\mathbf{A}$ of the protein interaction network (where the elements A_{ij} of the adjacency matrix equal $1/k_j$ if i and j are directly interacting, and equal 0 otherwise, with k_j being the degree of node j), the equation underlying this walk is given by

$$p_i^{(t+1)} = (1 - r) \sum_j A_{ij} p_j^{(t)} + r p_i^{(0)}, \qquad (9.6)$$

where $p_i^{(t)}$ is the probability that the walk arrives at protein i at time t, and starting probabilities $p_i^{(0)}$ are equal for all the disease-causing proteins. The second term on the right-hand-side of the above equation ensures that the walk remains confined to the vicinity of the disease-causing proteins and does not drift too far. Suppose the walk is declared to be converged at time T. The nodes in the protein interaction network are then ranked in decreasing order of $p_i^{(T)}$. It turns out that such global methods of search for proteins related to disease-causing proteins on a protein interaction network significantly outperform local methods based on direct interactions or shortest paths, especially for identifying disease genes for complex polygenic disorders and cancers.

The correspondence between phenotypic closeness and closeness along a protein interaction network was beautifully exploited to predict new disease genes for human inherited ataxias (Lim et al., 2006). Ataxias are neurodegenerative disorders characterized by loss of balance and coordination. To elucidate the local protein interaction network for ataxia, Lim et al. started with twenty-three known ataxia-causing proteins, and extended this list by incorporating thirty-one additional proteins that are known to interact with the twenty-three original proteins or are paralogs of these proteins. These fifty-four proteins were used as bait proteins for a yeast two-hybrid experiment to find all other proteins in the human proteome that interact with the bait proteins. This experiment revealed 770 interactions among the 54 baits and 561 identified preys. This network was further extended by incorporating manually curated interactions from the literature that involved either baits or preys and interactions inferred by orthology to other organisms. These efforts resulted in a large "ataxia network" consisting of 6,972 protein interactions among 3,607 proteins, such that every protein node in the network has either first-, second-, or third-order interactions with the original set of twenty-three ataxia-causing proteins. This network was mined in various ways to reveal insights into the common mechanisms behind various ataxias (where common mechanisms are revealed by direct or indirect protein-protein interactions between ataxia-causing proteins associated with different ataxias), and was also used to predict new ataxia-causing proteins based on their interactions with known ataxia-causing proteins.

Just as the list of ataxia-causing proteins was expanded by considering interacting proteins, entire disease pathways can be extended by taking into account the network neighborhood of the pathway under consideration. Earlier, we encountered signaling or signal transduction pathways, which are sets of proteins that sequentially transmit a signal from the cell surface to the nucleus (for this reason, a signaling pathway is also termed a *cascade* of proteins). Mutation of any one of the corresponding genes in the pathway can lead to a disease state because of lack of transmission of the signal. For example, many signals lead to inhibition of cell growth; thus disruption of the corresponding pathway can lead to cancer. Because disruption can occur by mutations in any of the genes in the pathway, all of those genes are disease genes. New disease-causing genes can be identified by extending pathways via systematic incorporation of proteins that interact with pathway proteins. For example, Glaab et al. (2010) augmented known disease pathways by adding proteins that satisfied certain conditions in terms of the number of "within-pathway" and "outside-pathway" interactions they have. This augmented network reveals a number of new disease genes.

9.3.2 Predicting Disease Genes *ab initio*

While the disease gene prediction methods just discussed employ information in protein interaction networks, they also rely on the availability of known disease genes. It is, however, possible to carry out new disease gene prediction without knowledge of other disease genes. These *ab initio* methods make use of the fact that disease genes occur as members of differentially expressed modules or complexes in protein interaction networks.

The use of differential expression to identify genes involved in disease actually predates the incorporation of protein interaction data. Using microarrays, it is possible to carry out genome-wide scans for mRNA expression in both normal and diseased states, and compare the two mRNA expression profiles to identify genes that are significantly differentially expressed in the two conditions. This process typically leads to a large number of identified genes, not all of which may be directly implicated in the disease. These latter genes are false positives. The microarray method is also capable of producing false negative errors: true disease genes that are not in fact differentially expressed but may differently regulate the expression of other genes or whose protein products may interact with other differentially expressed proteins. In comparison, the process of identifying disease genes by looking for them as members of differentially expressed protein interaction *modules* is far less error-prone.

Strategies that combine expression and network data to predict new disease genes mostly differ in the method by which a candidate protein is identified as one that is surrounded by differentially expressed proteins. Modern methods to identify these node neighborhoods use global information that includes indirect interactions and paths rather than direct interactions alone. For example, even if a protein node j does not directly interact with node i, there may be a very large number of shortest paths connecting j to i, thus making j a valid candidate to belong in the neighborhood of i. Methods to identify such global neighborhoods make use of a mathematical construct called a *kernel matrix* that can be computed from the adjacency matrix. Loosely speaking, a kernel matrix $\mathbf{K}$ on a network with n nodes is an $n \times n$ symmetric matrix that quantifies, in some sense, the closeness or similarity between two nodes in the network. A concrete example of a kernel matrix is the Laplacian exponential diffusion kernel defined below. In general, the kernel matrix can be defined in various ways, depending on the notion of closeness that is most relevant to the particular problem. A given kernel matrix can be used to define a distance function $d(i,j)$ between nodes i and j:

$$d(i,j) = K_{ii} + K_{jj} - 2K_{ij}. \tag{9.7}$$

This distance function can, in turn, be used to define the neighborhood of a gene/protein (as, say, all nodes within a certain distance of the protein of interest) in a protein interaction network. The proteins belonging to the neighborhood can then be examined for differential expression in a disease state as compared to the normal state. If the proteins in the neighborhood collectively exhibit significant differential expression, the protein in question is predicted to be a disease-causing protein. Thus, in these methods, knowledge of disease proteins is replaced by knowledge of expression levels.

As a concrete example, Nitsch et al. (2009) used as a starting point a human protein interaction network derived from the STRING database that we first encountered in Chapter 3 as a database that facilitates the prediction of protein-protein interactions using gene neighborhood analysis. From the derived or predicted human protein interaction network, Nitsch et al. computed the elements of a *Laplacian exponential diffusion* kernel matrix, defined as

$$\mathbf{K} = e^{\beta \mathbf{L}}, \tag{9.8}$$

where $\mathbf{L}$ is the Laplacian matrix of the network (see Box 8.2). They used the Laplacian exponential diffusion kernel to obtain a distance function between nodes, as stated above. Instead of explicitly defining node neighborhoods, they considered the entire network as belonging to the neighborhood of a node, but down-weighted the differential expression of distant nodes by their distance from the node in question. The score or propensity of a candidate protein to be a disease protein was defined as the maximum difference between the differential expression level of a neighborhood protein (after down-weighting its differential expression as mentioned) and randomized differential expression. High-scoring candidate proteins were predicted to cause disease.

The same research group later compared kernel-based approaches for network analysis of differential expression with a naive approach that relied on direct interactors alone (Nitsch et al., 2010), and found, in a validation study, that disease genes were ranked at an average ranking of 8 out of 100 in the best kernel-based approach, relative to an average ranking of 17 out of 100 when only direct interactors were included. This suggests that global positioning of genes relative to a disease gene in a protein interaction network is more informative of disease than local or direct interaction alone.

9.4 Network Analysis of Disease Mechanisms and Disease Response

We have discussed how disease proteins occupy special positions relative to each other in protein interaction networks—disease proteins that are causative of the same or similar diseases tend to cluster close to each other in these networks. This fact enables the prediction of new disease proteins, as was explained above. Moreover, it suggests that, in general, genetic diseases may be properly ascribed to protein interaction *modules* or clusters rather than to single proteins or genes. This way of thinking not only has consequences for the identification of new disease genes/proteins, but also for illuminating disease mechanisms by identifying new relationships among diseases, as we now proceed to explain.

In the beginning of this chapter, we pointed out that construction of the Human Disease Network, in which two diseases are linked if they share at least one causative disease gene, led to new insights about the relationships between diseases. However, even more interesting relationships can be unearthed by the simultaneous incorporation of the protein interaction network into this paradigm. Consider, for instance, two diseases A and B that may not ordinarily be linked in the HDN because they do not share any common disease gene. Suppose that one of the disease proteins for A has a protein-protein interaction with one of the disease proteins for B. Because closeness on a protein interaction network is strongly correlated with phenotypic closeness, this suggests that diseases A and B have a common mechanism. Incorporation of protein interaction data thus provides higher-order structure to the HDN. For example, by constructing a large protein interaction network via data integration from disparate sources, Wu et al. (2010) showed that cancer genes were clustered in a number of network modules that are common to different cancers. One of the modules found corresponds to proteins involved in cell cycle regulation, DNA repair, and other nuclear processes, while a second common module corresponds to signal transduction events in the plasma membrane and the cytoplasm. This suggests that the development of cancerous cells generally requires mutations in both nuclear and cytoplasmic/plasma membrane-based pathways.

Other studies that throw light on the evolution of disease states focus on the global positioning of disease proteins in protein interaction networks. Goh et al. (2007) classified human disease genes into "essential" and "nonessential" categories, where essential genes were those whose disruption would result in embryonic or postnatal lethality, that is, the individual hosting the disrupted gene would die before attaining maturity. Interestingly, and in analogy with baker's yeast, they found that essential proteins were enriched for hubs, while nonessential disease proteins showed no such tendency. Furthermore, essential genes were found to display high average co-expression with other genes while nonessential disease genes showed the opposite effect, having expression patterns that either uncorrelated or anti-correlated with other genes. These differences between essential and nonessential disease proteins can be explained by the fact that mutations in topologically and functionally central proteins are more likely to lead to lethality than mutations in peripheral proteins. Thus, nonessential disease proteins evolve to occupy peripheral positions (i.e., have low degree and low centrality) in interaction networks so that mutations in these proteins can still allow the host cell to be viable, because disease genes can be inherited only if the person harboring the disease can survive to maturity and be viable.

The above argument does not hold true for diseases that result from *somatic* mutations. Somatic mutations are mutations that occur in the genomes of non-germ cells (in humans, the sperm and egg cells are the germ cells) and are therefore not transmitted to the next generation. Disease genes that undergo somatic mutations are not under evolutionary pressure to code for proteins that are positioned at the peripheries of protein interaction networks. Indeed, at least two independent studies have found that somatic cancer genes are more likely to encode hubs and more likely to show high co-expression with all other genes in the cell (Jonsson and Bates, 2006; Goh et al., 2007).

There are other diseases that have a large number of disease genes in the sense that the DNA sequences of these genes are altered in at least some members of the population carrying the disease; however, the set of these genes varies across different populations so that no one gene or one set of genes can be said to be a disease gene/set for the disease in all individuals. These genes are not true disease genes because they are not surely implicated in every occurrence of the disease but specific sequence variants of these genes may constitute "risk" factors for certain diseases. How does network biology impact our understanding of such diseases, and how does one get around the fact that the concept of the disease gene is somewhat "fuzzy" in these cases? One strategy that we encountered in the previous section is to replace the notion of a disease gene by that of a gene that is differentially expressed between normal and diseased individuals. In other words, the gene itself may not be mutated but its expression level might be altered, perhaps due to epigenetic modifications.

One example of a disease of the above type is asthma, which is a complex polygenic disease that is also *multifactorial*, that is, it arises out of the interaction between genetic propensity and environmental factors. The subject of identifying disease genes associated with asthma is a field of ongoing research, with new disease genes still being discovered in different populations. Rather than dealing with a dynamic set of putative disease genes, it perhaps makes more sense to identify the genes that are differentially expressed in a specific instance of the disease state and then analyze the network properties of the differentially expressed genes. This is exactly what Lu et al. (2007) did in a network analysis of differentially expressed genes in asthmatic mice.

Lu et al. (2007) studied two types of mice, one wild-type and another that was deficient in recombinase-activating genes (RAG). These genes are responsible for generating the wide spectrum of antibodies in response to foreign antigens. When the RAG genes are knocked out (mutated), the immune system does not generate the requisite antibodies. Thus, when these two types of mice were exposed to an asthma-inducing allergen such as ovalbumin, wild-type mice generated an immune response characterized by hyper-responsiveness of the airway, increased serum Immunoglobulin E levels, and other such signatures of the asthmatic response. On the other hand, RAG-deficient mice generated no such immune response. RNA extracts from the lung tissue of these two mice types were than analyzed by microarrays to identify over 700 genes that are differentially expressed among the two types of mice.

When analyzing the protein interaction network properties of the proteins corresponding to these 700 differentially expressed genes, Lu et al. (2007) found that hub proteins (defined in this study as proteins with degree >5) exhibited very low differential expression as compared to peripheral proteins that were far more highly differentially expressed. Furthermore, when the network of shortest paths connecting hub proteins was analyzed, it was found that this network contained "superhubs": proteins with degree >5 in the shortest path network; these superhubs had even lower differential expression than the other hubs. These findings are consistent with the idea that high differential expression in hubs and superhubs would be disruptive across the entire protein interaction network, and exposure to allergen would therefore cause greater harm to the individual. By relegating high differential expression to peripheral proteins, the immune response is more manageable. This study paves the way for the prediction of gene expression modulation in immune response to allergens purely by analysis of the network connectivity. Moreover, it shows that hub proteins that have low, albeit statistically significant level of modulation may be completely missed by expression analysis protocols that look for large changes in expression, even though these proteins might well have an important role in the disease.

While differential expression of individual genes yields insight into disease mechanisms when these genes are overlayed on interaction networks, analysis of the global patterns of differential expression is also revealing. For example, similar to the dynamic modularity encountered in the yeast interactome (as discussed in the previous chapter), the human protein interaction network also contains two kinds of hub proteins: those that have strong co-expression with their interacting partners (*intramodular* hubs or party hubs), and those that have weak co-expression with their interacting partners (*intermodular* hubs or date hubs). Because we know that cancer proteins occupy highly central positions in protein interaction networks, it is not surprising that these proteins have higher propensity to be intermodular hubs than intramodular ones. The division of hub proteins into two classes depends crucially on the global expression pattern of yeast proteins. Because the expression pattern changes during disease progression, one would expect this division also to change. The manner in which this classification is altered during disease progression turns out to be revealing.

A specific example of this kind of alteration is breast cancer, in which hubs whose correlation of expression with their interacting partners varies significantly, depending upon the nature of progression of the disease, can be readily identified. Thus, one can pinpoint hubs whose correlated expression is significantly different in breast cancer patients who eventually became disease-free and survived ("good outcome" patients) relative to patients who died of the disease ("poor outcome" patients). In an analysis of this type carried out by Taylor et al. (2009), 256 hubs were identified whose correlation of expression with their interacting partners was significantly different between good and poor outcome patients. Loosely speaking, these 256 hubs have different identities in the two groups of patients: intramodular hubs in one group are intermodular in the other, and vice versa. The expression correlation of these hubs is therefore a signature of disease outcome and can be used to predict prognosis. Remarkably, prediction accuracy for the probability of ten-year survival using this method (about 71%) compares favorably with that using commercially available genomic breast cancer diagnostic kits, and presumably this accuracy would only increase if the methods were to be combined in some way.

Even the expression correlation of certain individual genes is highly predictive of breast cancer outcome. For example, the protein coded by the gene $BRCA1$ is an intramodular hub in tumors from patients who eventually survive but an intermodular hub in tumors from poor outcome patients.

9.5 Network-Based Prediction of Disease Comorbidity

Toward the beginning of this chapter, we discussed the construction of the HDN, in which two diseases are linked if they share at least one common disease gene, and we stated that diseases in the HDN are clustered naturally according to disorder classes. This fact suggests that grouping diseases by the common identity of the genes that cause them uncovers the natural phenotypic groups to which these diseases belong. One can probe the functional relevance of the HDN a bit further and ask to what extent, and in what manner, the HDN clarifies functional relationships between diseases. A physiologically important quantity is the extent to which two diseases will tend to co-occur in the same individual. This propensity is termed the *comorbidity* of the two diseases, and it is natural to ask whether connections in the HDN reveal comorbidity patterns.

While comorbidity between a pair of diseases can be quantified in various ways, it is most simply defined as the frequency with which the two diseases actually occur within the same individual over a certain time period, divided by the frequency with which they would be expected to occur by chance in the same time period. The *incidence* of a disease is the expected number of new cases diagnosed with the disease over a specified period of time, divided by the population that is at risk for the disease. Thus, the incidence I_i is the expected frequency of occurrence of new cases of disease i among the population within a specified time period. A simple measure of comorbidity can then be expressed as

$$C = \frac{f_{ij}}{I_i I_j},$$
(9.9)

where f_{ij} is the proportion of members of the population who develop both disease i and disease j within the time period studied. Note that $C > 1$ would imply that the two diseases in question are more likely to occur in the same individual than can be explained by chance alone.

Returning to the relationship between comorbidity and the HDN, it is indeed found that diseases that are directly linked in the HDN have average comorbidity that is two- to four-fold higher than the average comorbidity between all pairs of diseases (Park et al., 2009) (see Table 9.1), and that furthermore, the average comorbidity increases with the number of shared disease genes among the two diseases in question. A possible explanation for this effect is as follows: two diseases are linked in the HDN if they have the same disease gene, that is, mutations in the same gene can cause both diseases. Thus, an individual who has one of the diseases likely has a mutated version of the gene causing the disease, which in turn is likely to cause the other disease to also occur. However, by this line of reasoning, the high observed comorbidity among diseases linked in the HDN should not occur if the mutations that cause them lie on separate functional parts of the gene, or equivalently, on separate domains of the corresponding protein. In such cases, an individual who suffers from one of the diseases harbors a protein with one disrupted functional domain, but with the rest of the domains perfectly functional. Such an individual does not have enhanced propensity to develop the other disease. As shown in Table 9.1, this line of reasoning also holds true: disease pairs that are linked in the HDN and have causative mutations that lie on different protein domains ("non-domain sharing") have lower average comorbidity than the overall average comorbidity of linked diseases, while those whose causative mutations lie on the same domain ("domain sharing") have higher average comorbidity than the overall average comorbidity of linked diseases.

Is linkage or "closeness" in the HDN the best predictor of disease comorbidity? Strictly speaking, this is an open question at the time of writing, although Park et al. (2009) compared linkage in the HDN with two other predictors of comorbidity, namely the existence of a protein-protein interaction between two proteins that are respectively causative of the two diseases, and correlation in expression of the disease-causing

TABLE 9.1:   Average Comorbidity of Disease Pairs Satisfying Various Criteria

Criteria	Number of Pairs	Average Comorbidity
All diseases	83, 924	1.92
Linked in the HDN	658	4.35
Domain sharing, linked in the HDN	182	5.98
Non-domain sharing, linked in the HDN	476	3.73
PP interaction	1, 873	2.35
Correlated expression	215	2.79

Source: Park et al. (2009).

genes of the two diseases. It turns out that neither of these two features is a very good predictor of comorbidity: Table 9.1 shows that the average comorbidity of disease pairs that have underlying protein-protein interactions or correlated expression is not appreciably higher than the average comorbidity of all pairs of diseases. However, the predictive power of protein-protein interactions for comorbidity is significantly enhanced when coupled with protein localization information: two diseases whose causative proteins are "close" in the human protein interaction network and localized in the same cellular compartment are likely to display high levels of comorbidity (Park et al., 2011). Interestingly, the fact that the average comorbidity of all disease pairs is larger than 1 reflects the observation that most people who suffer from at least one disease are likely to develop several other afflictions.

One problem with using linkage in the HDN to infer comorbidity arises due to the fact that the most disconnected class of diseases in the HDN are the metabolic diseases. Thus, direct linkage in the HDN is not a good predictor of comorbidity among metabolic diseases. However, it may be that some type of higher-order linkage correlates well with comorbidity. For example, Lee et al. (2008a) found that two metabolic diseases caused by loss of function of metabolic enzymes that catalyze *adjacent* reactions (adjacent reactions process a common metabolite) in a metabolic pathway are likely to have high comorbidity. They constructed a metabolic disease network such that two diseases in this network are linked if their causative enzymes process the same metabolite, and further found that the propensity for comorbidity of two diseases increases as diseases get closer to each other in the metabolic disease network. Studies such as this suggest that deeper mining of disease networks should be fruitful for the elucidation of disease comorbidity.

9.6 Cancer and Synthetic Lethality

One of the most promising applications of the discovery of synthetic lethal relationships between genes is toward novel strategies for the treatment of cancer. The genetic basis of cancer suggests that, in theory, it should be possible to treat cancer by designing drugs that target known disease genes for cancer. This strategy fails in practice because of the side effects associated with the artificial regulation of the same genes in normal cells, and also because many cancer-related proteins are "undruggable" for very simple reasons. Drugs that are ultimately approved for medicinal use fall into two major structural classes: either they are small molecules that require the presence of a "hydrophobic pocket" in a protein in order to bind to it, or they are "biologics," that is, proteins that require a relatively flat contact surface (not necessarily highly hydrophobic) on another protein in order to bind to it. Biologics, however, have a limitation: because of their large size, they cannot traverse mammalian cell membranes and can therefore only target extracellular proteins. To summarize, these characteristics of available drug types imply that a "druggable" protein must be extracellular or possess a hydrophobic pocket (or both). Unfortunately, of the ~25,000 human proteins known, only about 3,000 possess a hydrophobic pocket that is suitable for tight binding of small molecules, and less than 10% are extracellular. This leaves 75% to 80% of human proteins beyond the reach of established types of drugs. Unfortunately, most cancer proteins fall squarely within this large class: they are expressed within the cell and, because they function primarily via protein-protein interactions, possess extended flat surfaces but no hydrophobic pocket.

These are some of the challenges that underlie putative therapies for cancer, but they are potentially resolvable, at least in theory, by the exploitation of synthetic lethal relationships between cancer genes and other genes. Suppose we know of a cancer gene C, that is, a gene that exists in mutated form in cancerous cells, and also another gene S that has a synthetic lethal relationship with C, that is, human cells function normally when either S or C loses function, but die when they both lose function. Suppose further that the protein coded by S is druggable. Then a drug that targets the protein coded by S will cause loss of function in S in every cell. Because C has not lost function in normal cells, normal cells will not be affected by the drug. However, because C has lost function in a cancerous cell, the simultaneous loss of function of S effected by the drug in the cell will cause the cancerous cell to die. This potential treatment for cancer therefore does not require the protein product of C to be druggable (although the protein product of S must be), and should not, in theory, cause side effects because normal cells will remain unaffected by the drug. This argument also holds for C that has a gain-of-function or a dominant mutation as long as the mutant gene product is restricted to the cancer cells.

Certain breast cancer types that are highly resistant to conventional treatment strategies such as chemotherapy and radiation include those that have a mutated form of the *BRCA1* or *BRCA2* gene, both of which produce proteins that help repair damaged DNA. A separate gene *PARP1* also has a protein product with the same function and has a synthetic lethal relationship with these genes. Thus, drugs that target *PARP1* offer attractive lines of treatment for these cancers and have led to promising results in clinical trials. Similarly, the discovery of synthetic lethal relationships between the epidermal growth factor receptor EGFR (a protein that is overexpressed in a number of cancers) and other genes using RNAi-based screens (see discussion in Section 2.6) could be used to design a therapy that simultaneously targets EGFR and its synthetic lethal partners (Astsaturov et al., 2010). In another example, the oncogene *KRAS* is mutated in a number of cancers, including pancreatic, lung, and colon, but the protein coded by it is undruggable. The discovery of at least three different enzymes that have synthetic lethal relationships with *KRAS* should lead to promising treatment strategies (Scholl et al., 2009; Barbie et al., 2009; Luo et al., 2009). An examination of the synthetic lethal interaction between the *KRAS* gene and nearly half of all identified genes was carried out using an inhibitory RNA screen on cancer cells (Luo et al., 2009). The potential therapeutic value of synthetic lethality is therefore expected to inspire the development of new computational and experimental methodologies to uncover human synthetic lethal networks at the genome scale. Such methods will increasingly be used to develop the repertoire of novel approaches to cancer therapies.

References

Albert, R. and Barabási, A.-L. Statistical mechanics of complex networks. *Reviews of Modern Physics* 74:47 (2002).

Albert, R. and Othmer, H. G. The topology of the regulatory interactions predicts the expression pattern of the *Drosophila* segment polarity genes. *Journal of Theoretical Biology* 223:1 (2003).

Aldana, M. and Cluzel, P. A natural class of robust networks. *Proceedings of the National Academy of Sciences USA* 100:8710 (2003).

Aloy, P. and Russell, R. B. Interrogating protein interaction networks through structural biology. *Proceedings of the National Academy of Sciences USA* 99:5896 (2002).

Ashburner, M., Ball, C. A., Blake, J. A., et al. Gene ontology: tool for the unification of biology. *Nature Genetics* 25:25 (2000).

Astsaturov, I., Ratushny, V., Sukhanova, A., et al. Synthetic lethal screen of an *EGFR*-centered network to improve targeted therapies. *Science Signaling* 3:ra67 (2010).

Avery, L. and Wasserman, S. Ordering gene function: The interpretation of epistasis in regulatory hierarchies. *Trends in Genetics* 8:312 (1992).

Bader, G. D. and Hogue, C. W. V. Analyzing protein-protein interaction data obtained from different sources. *Nature Biotechnology* 20:991 (2002).

Bailey, J., Gu, Z., Clark, R., Reinert, K., and Samonte, R. Recent segmental duplications in the human genome. *Science* 297:1003 (2002).

Bandyopadhyay, S., Mehta, M., Kuo, D., et al. Rewiring of genetic networks in response to DNA damage. *Science* 330:1385 (2010).

Barabási, A.-L. Network medicine — from obesity to the "diseasome." *New England Journal of Medicine* 357:404 (2007).

Barabási, A.-L. and Albert, R. Emergence of scaling in random networks. *Science* 286:509 (1999).

Barbie, D. A., Tamayo, P., Boehm, J. S., et al. Systematic RNA interference reveals that oncogenic *KRAS*-driven cancers require *TBK1*. *Nature* 462:108 (2009).

Basso, K., Margolin, A., Stolovitzky, G., et al. Reverse engineering of regulatory networks in human B cells. *Nature Genetics* 37:382 (2005).

Beal, M. J., Falciani, F., Ghahramani, Z., Rangel, C., and Wild, D. L. A Bayesian approach to reconstructing genetic regulatory networks with hidden factors. *Bioinformatics* 21:349 (2005).

Becker, S. A., Feist, A. M., Mo, M. L., et al. Quantitative prediction of cellular metabolism with constraint-based models: The COBRA toolbox. *Nature Protocols* 2:727 (2007).

Blanc, G., Barakat, A., Guyot, R., Cooke, R., and Delseny, M. Extensive duplication and reshuffling in the *Arabidopsis* genome. *Plant Cell* 12:1093 (2000).

Bollobás, B. *Random Graphs*. Cambridge University Press, Cambridge, 2nd ed. (2001).

Carter, G. W., Galas, D. J., and Galitski, T. Maximal extraction of biological information from genetic interaction data. *PLoS Computational Biology* 5:e10000347 (2009).

Chen, K. C., Calzone, L., Csikasz-Nagy, A., et al. Integrative analysis of cell cycle control in budding yeast. *Molecular Biology of the Cell* 15:3841 (2004).

Chung, F. and Lu, L. *Complex Graphs and Networks*. American Mathematical Society, Providence, RI (2006).

Collins, S. R., Schuldiner, M., Krogan, N. J., and Weismann, J. S. A strategy for extracting and analyzing large-scale quantitative epistatic interaction data. *Genome Biology* 7:R63 (2006).

Cooper, G. M. *The Cell: A Molecular Approach*. Sunderland, MA: Sinauer Associates, 2nd ed. (2000).

Cordell, H. J. Epistasis: What it means, what it doesn't mean, and statistical methods to detect it in humans. *Human Molecular Genetics* 11:2463 (2002).

Court, D. L., Oppenheim, A. B., and Adhya, S. L. A new look at bacteriophage λ genetic networks. *Journal of Bacteriology* 189:298 (2007).

Covert, M. W., Schilling, C. H., and Palsson, B. Regulation of gene expression in flux balance models of metabolism. *Journal of Theoretical Biology* 213:73 (2001).

Dandekar, T., Snel, B., Huynen, M., and Bork, P. Conservation of gene order: A fingerprint of proteins that physically interact. *Trends in Biochemical Sciences* 23:324 (1998).

Dartnell, L., Simeonidis, E., Hubank, M., et al. Robustness of the *p53* network and biological hackers. *FEBS Letters* 579:3037 (2005).

Deane, C., Salwinski, L., Xenarios, I., and Eisenberg, D. Protein interactions: Two methods for assessment of the reliability of high throughput observations. *Molecular Cell Proteomics* 1:349 (2002).

Deeds, E., Ashenberg, O., and Shakhnovich, E. A simple physical model for scaling in protein-protein interaction networks. *Proceedings of the National Academy of Sciences USA* 103:311 (2006).

Delbrück, M. A physicist looks at biology. In *Phage and the Origins of Molecular Biology*, pp. 9. Cold Spring Harbor, NY (1966).

Dempster, A. P., Laird, N. M., and Rubin, D. B. Maximum likelihood estimation from incomplete data via the EM algorithm. *Journal of the Royal Statistical Society, Series B* 39:1 (1977).

Deng, M., Mehta, S., Sun, F., and Chen, T. Inferring domain-domain interactions from protein-protein interactions. *Genome Research* 12:1540 (2002).

d'Haeseleer, P. and Church, G. M. Estimating and improving protein interaction error rates. In *Computational Systems Bioinformatics (CSB)*, pp. 216–223 (2004).

Dodd, I. B., Shearwin, K. E., and Egan, J. B. Revisited gene regulation in bacteriophage λ. *Current Opinion in Genetics and Development* 15:145 (2005).

Dojer, N., Gambin, A., Mizera, A., Wilczyński, B., and Tiuryn, J. Applying dynamic Bayesian networks to perturbed gene expression data. *BMC Bioinformatics* 7:249 (2006).

Dorogovtsev, S., Mendes, J., and Samukhin, A. Structure of growing networks with preferential linking. *Physical Review Letters* 85:4633 (2000).

Dunn, R., Dudbridge, F., and Sanderson, C. M. The use of edge-betweenness clustering to investigate biological function in protein interaction networks. *BMC Bioinformatics* 6:39 (2005).

Erdös, P. and Renyi, A. On random graphs. i. *Publicationes Mathematicae* 6:290 (1959).

Erdös, P. and Renyi, A. On the evolution of random graphs. *Publication of the Mathematical Institute of the Hungarian Academy of Sciences* 5:17 (1960).

Evlampiev, K. and Isambert, H. Modeling protein network evolution under genome duplication and domain shuffling. *BMC Systems Biology* 1:49 (2007).

Evlampiev, K. and Isambert, H. Conservation and topology of protein interaction networks under duplication-divergence evolution. *Proceedings of the National Academy of Sciences USA* 105:9863 (2008).

Faith, J. J., Hayete, B., Thaden, J. T., Mogno, I., and Wierzbowski, J. Large-scale mapping and validation of *escherichia coli* transcriptional regulation from a compendium of expression profiles. *PLoS Biology* 5:e8 (2007).

Fernandez, A. Molecular basis for evolving modularity in the yeast protein interaction network. *PLoS Computational Biology* 3:e226 (2007).

Ferrell, J. E., Jr. Self-perpetuating states in signal transduction: Positive feedback, double-negative feedback and bistability. *Current Opinion in Cell Biology* 14:140 (2002).

Fishel, R., Lescoe, M. K., Rao, M. R. S., et al. The human mutator gene homolog *MSH2* and its association with hereditary nonpolyposis colon cancer. *Cell* 75:1027 (1993).

Force, A., Lynch, M., Pickett, F., et al. The preservation of duplicate genes by complementary degenerative mutations. *Genetics* 151:1531 (1999).

Formstecher, E., Aresta, S., Collura, V., Hamburger, A., and Meil,

A. Protein interaction mapping: A *Drosophila* case study. *Genome Research* 15:376 (2005).

Fortunato, S. Community detection in graphs. *Physics Reports* 486:75 (2010).

Fraser, H. B. and Plotkin, J. B. Using protein complexes to predict phenotypic effects of gene mutation. *Genome Biology* 8:R252 (2007).

Friedman, N. Learning belief networks in the presence of missing values and hidden variables. In *Proceedings of the 14th International Conference on Machine Learning*, pp. 125–133. Nashville, TN (1997).

Friedman, N., Murphy, K., and Russell, S. Learning the structure of dynamic probabilistic networks. In *Proceedings of the Fourteenth Conf. on Uncertainty in Artificial Intelligence (UAI)*, pp. 139–147. Morgan Kaufmann (1998).

Fuhrman, S. and Somogyi, R. Distributivity, a general information theoretic network measure, or why the whole is more than the sum of its parts. In *Proceedings of the International Workshop on Information Processing in Cells and Tissues (IPCAT)* (1997).

Gavin, A.-C., Bösche, M., Krause, R., et al. Functional organization of the yeast proteome by systematic analysis of protein complexes. *Nature* 415:141 (2002).

Gevers, D., Vandepoele, K., Simillion, C., and van de Peer, Y. Gene duplication and biased functional retention of paralogs in bacterial genomes. *Trends in Microbiology* 12:148 (2004).

Giacomantonio, C. E. and Goodhill, G. J. A Boolean model of the gene regulatory network underlying mammalian cortical area development. *PLoS Computational Biology* 6:e1000936 (2010).

Gilbert, S. F. *Developmental Biology*. Sinauer Associates, Sunderland, MA, 6th ed. (2000).

Glaab, E., Baudot, A., Krasnogor, N., and Valencia, A. Extending pathways and processes using molecular interaction networks to analyse cancer genome data. *BMC Bioinformatics* 11:597 (2010).

Goh, K.-I., Cusick, M. E., Valle, D., et al. The human disease network. *Proceedings of the National Academy of Sciences USA* 104:8685 (2007).

Goldberg, D. S. and Roth, F. P. Assessing experimentally derived interactions in a small world. *Proceedings of the National Academy of Sciences USA* 100:4372 (2003).

Gray, M. W., Burger, G., and Lang, B. F. Mitochondrial evolution. *Science* 283:1476 (1999).

Green, R. A., Kao, H.-L., Audhya, A., et al. A high-resolution *C. elegans* essential gene network based on phenotypic profiling of a complex tissue. *Cell* 145:470 (2011).

Guimarães, K. S., Jothi, R., Zotenko, E., and Przytycka, T. M. Predicting domain-domain interactions using a parsimony approach. *Genome Biology* 7:R104 (2006).

Guimera, R. and Amaral, L. A. N. Functional cartography of complex metabolic networks. *Nature* 433:895 (2005).

Guimera, R., Sales-Pardo, M., and Amaral, L. A. N. Module identification in bipartite and directed networks. *Physical Review E* 76:036102 (2007).

Hahn, M. W. and Kern, A. D. Comparative genomics of centrality and essentiality in three eukaryotic protein-interaction networks. *Molecular Biology and Evolution* 22:803 (2005).

Han, J., Bertin, N., Hao, T., Goldberg, D., and Berriz, G. Evidence for dynamically organized modularity in the yeast protein-protein interaction network. *Nature* 430:88 (2004).

Hart, G. T., Ramani, A. K., and Marcotte, E. M. How complete are current yeast and human protein-interaction networks? *Genome Biology* 7:120 (2006).

Herrgard, M. J., Swainston, N., Dobson, P., et al. A consensus yeast metabolic network reconstruction obtained from a community approach to systems biology. *Nature Biotechnology* 26:1155 (2008).

Ho, Y., Gruhler, A., Heilbut, A., et al. Systematic identification of protein complexes in *Saccharomyces cerevisiae* by mass spectrometry. *Nature* 415:123 (2002).

Huang, H., Jedynak, B. M., and Bader, J. S. Where have all the interactions gone? Estimating the coverage of two-hybrid protein interaction maps. *PLoS Computational Biology* 3:e214 (2007).

Husmeier, D. Sensitivity and specificity of inferring genetic regulatory interactions from microarray experiments with dynamic Bayesian networks. *Bioinformatics* 19:2271 (2003).

Iliopoulos, I., Enright, A. J., Poullet, P., and Ouzounis, C. A. Mapping functional associations in the entire genome of *Drosophila melanogaster* using fusion analysis. *Comparative and Functional Genomics* 4:337 (2003).

Ingolia, N. T. Topology and robustness in the *Drosophila* segment polarity network. *PLoS Biology* 2:e123 (2004).

Ito, T., Chiba, T., Ozawa, R., et al. A comprehensive two-hybrid analysis to explore the yeast protein interactome. *Proceedings of the National Academy of Sciences USA* 98:4569 (2001).

Jansen, R., Yu, H., Greenbaum, D., et al. A Bayesian networks approach for predicting protein-protein interactions from genomic data. *Science* 302:449 (2003).

Jeong, H., Mason, S., Barabási, A.-L., and Oltvai, Z. Lethality and centrality in protein networks. *Nature* 411:41 (2001).

Jeong, H., Tombor, B., Albert, R., Oltvai, Z., and Barabási, A.-L. The large-scale organization of metabolic networks. *Nature* 407:651 (2000).

Jones, S. and Thornton, J. M. Analysis of protein-protein interaction sites using surface patches. *Journal of Molecular Biology* 272:121 (1997a).

Jones, S. and Thornton, J. M. Prediction of protein-protein interaction sites using patch analysis. *Journal of Molecular Biology* 272:133 (1997b).

Jonsson, P. F. and Bates, P. A. Global topological features of cancer proteins in the human interactome. *Bioinformatics* 22:2291 (2006).

Jothi, R., Cherukuri, P. F., Tasneem, A., and Przytycka, T. M. Co-evolutionary analysis of domains in interacting proteins reveals insights into domain-domain interactions mediating protein-protein interactions. *Journal of Molecular Biology* 362:861 (2006).

Juan, D., Pazos, F., and Valencia, A. High-confidence prediction of global interactomes based on genome-wide coevolutionary networks. *Proceedings of the National Academy of Sciences USA* 105:934 (2008).

Kaelin, W. G., Jr. The concept of synthetic lethality in the context of anticancer therapy. *Nature Reviews Cancer* 5:689 (2005).

Kaizu, K., Ghosh, S., Matsuoka, Y., et al. A comprehensive molecular interaction map of the budding yeast cell cycle. *Molecular Systems Biology* 6:415 (2010).

Kanehisa, M., Araki, M., Goto, S., et al. KEGG for linking genomes to life and the environment. *Nucleic Acids Research* 36:480 (2008).

Keller, E. Revisiting "scale-free" networks. *BioEssays* 27:1060 (2005).

Kelley, R. and Ideker, T. Systematic interpretation of genetic interactions using protein networks. *Nature Biotechnology* 23:561 (2005).

Kellis, M., Birren, B., and Lander, E. Proof and evolutionary analysis of ancient genome duplication in the yeast *Saccharomyces cerevisiae*. *Nature* 428:617 (2004).

Kim, I., Liu, Y., and Zhao, H. Bayesian methods for predicting interacting protein pairs using domain information. *Biometrics* 63:824 (2007).

Kim, S., Imoto, S., and Miyano, S. Dynamic Bayesian network and nonparametric regression for nonlinear modeling of gene networks from time series gene expression data. *BioSystems* 75:57 (2004).

Kim, W. K. and Marcotte, E. M. Age-dependent evolution of the yeast protein interaction network suggests a limited role of gene duplication and divergence. *PLoS Computational Biology* 4:e1000232 (2008).

King, O. D. Comment on "Subgraphs in random networks." *Physical Review E* 70:058101 (2004).

Kitano, H. and Oda, K. Self-extending symbiosis: A mechanism for increasing robustness through evolution. *Biological Theory* 1:61 (2006).

Köhler, S., Bauer, S., Horn, D., and Robinson, P. N. Walking the interactome for prioritization of candidate disease genes. *American Journal of Human Genetics* 82:949 (2008).

Krastev, D. B., Slabicki, M., Paszkowski-Rogacz, M., et al. A systematic RNAi synthetic interaction screen reveals a link between *p53* and *snoRNP* assembly. *Nature Cell Biology* 13:809 (2011).

Krogan, N. J., Cagney, G., Yu, H., et al. Global landscape of protein complexes in the yeast *Saccharomyces cerevisiae*. *Nature* 440:637 (2006).

Kuntzer, J., Backes, C., Blum, T., et al. BNDB – The biochemical network database. *BMC Bioinformatics* 8:367 (2007).

Larhammer, D., Lundin, L., and Hallbook, F. The human *Hox*-bearing chromosome regions did arise by block or chromosome (or even genome) duplications. *Genome Research* 12:1910 (2002).

Lauritzen, S. L. The EM algorithm for graphical association models with missing data. *Computational Statistics and Data Analysis* 19:191 (1995).

Lawrence, M. C. and Colman, P. M. Shape complementarity at protein/protein interfaces. *Journal of Molecular Biology* 234:946 (2003).

Lee, D.-S., Park, J., Kay, K. A., et al. The implications of human metabolic network topology for disease comorbidity. *Proceedings of the National Academy of Sciences USA* 105:9880 (2008a).

Lee, H., Deng, M., Sun, F., and Chen, T. An integrated approach to the prediction of domain-domain interactions. *BMC Bioinformatics* 7:269 (2006).

Lee, J. M., Gianchandani, E. P., Eddy, J. A., and Papin, J. A. Dynamic analysis of integrated signaling, metabolic and regulatory networks. *PLoS Computational Biology* 4:e1000086 (2008b).

LeMeur, N. and Gentleman, R. Modeling synthetic lethality. *Genome Biology* 9:10 (2008).

Li, F., Long, T., Lu, Y., Ouyang, Q., and Tang, C. The yeast cell-cycle network is robustly designed. *Proceedings of the National Academy of Sciences USA* 101:4781 (2004).

Liang, S., Fuhrman, S., and Somogyi, R. REVEAL, a general reverse engineering algorithm for inference of genetic network architectures. *Pacific Symposium in Biocomputing* 3:18 (1998).

Lim, J., Hao, T., Shaw, C., et al. A protein-protein interaction network for human inherited ataxias and disorders of Purkinje cell degeneration. *Cell* 125:801 (2006).

Lin, N., Wu, B., Jansen, R., Gerstein, M., and Zhao, H. Information assessment on predicting protein-protein interactions. *BMC Bioinformatics* 5:154 (2004).

Lin, Y. C., Jhunjhunwala, S., Benner, C., et al. A global network of transcription factors, involving *E2A*, *EBF1* and *Foxo1*, that orchestrates B cell fate. *Nature Immunology* 11:635 (2010).

Little, J. W. and Shepley, D. P. Integrated analysis of multiple data sources reveals modular structure of biological networks. *Biochemical and Biophysical Research Communications* 345:302 (2006).

Little, J. W., Shepley, D. P., and Wert, D. W. Robustness of a gene regulatory circuit. *EMBO Journal* 18:4299 (1999).

Liu, Y., Kim, I., and Zhao, H. Protein interaction predictions from diverse sources. *Drug Discovery Today* 13:409 (2008).

Liu, Y., Liu, N., and Zhao, H. Inferring protein-protein interactions through high-throughput data from diverse organisms. *Bioinformatics* 21:3279 (2005).

Lu, H., Shi, B., Wu, G., et al. Integrated analysis of multiple data sources reveals modular structure of biological networks. *Biochemical and Biophysical Research Communications* 345:302 (2006).

Lu, X., Jain, V., Finn, P. W., and Perkins, D. L. Hubs in biological interaction networks exhibit low changes in expression in experimental asthma. *Molecular Systems Biology* 3:98 (2007).

Luo, J., Emanuele, M. J., Li, D., et al. A genome-wide RNAi screen identifies multiple synthetic lethal interactions with the *Ras* oncogene. *Cell* 137:835 (2009).

Lynch, M. and Conery, J. S. The evolutionary fate and consequences of duplicate genes. *Science* 290:1151 (2000).

Lynch, M. and Force, A. The probability of duplicate-gene preservation by subfunctionalization. *Genetics* 154:459 (2000).

Ma, W., Lai, L., Ouyang, Q., and Tang, C. Robustness and modular design of the *Drosophila* segment polarity network. *Molecular Systems Biology* 2:70 (2006).

Margolin, A. A., Basso, K., Wiggins, C., et al. ARACNE: An algorithm for the reconstruction of gene regulatory networks in a mammalian cellular context. *BMC Bioinformatics* 7:S7 (2006).

Markson, G., Kiel, C., Hyde, R., et al. Analysis of the human E2 ubiquitin conjugating enzyme protein interaction network. *Genome Research* 19:1905 (2009).

Matthews, L. R., Vaglio, P., Reboul, J., et al. Identification of potential interaction networks using sequence-based searches for conserved protein-protein interactions or "interologs." *Genome Research* 11:2120 (2001).

McLysaght, A., Hokamp, K., and Wolfe, K. Extensive genomic duplication during early chordate evolution. *Nature Genetics* 31:200 (2002).

Middendorf, M., Ziv, E., and Wiggins, C. Inferring network mechanisms: The *Drosophila melanogaster* protein interaction network. *Proceedings of the National Academy of Sciences USA* 102:3192 (2005).

Miller, G. *Information Theory in Psychology II-B*, chap. Note on the bias of information estimates, pp. 95–100. Glencoe, IL: Free Press (1955).

Milo, R., Shen-Orr, S., Itzkovitz, S., et al. Network motifs: Simple building blocks of complex networks. *Science* 298:824 (2002).

Molloy, M. and Reed, B. A critical point for random graphs with a given degree sequence. *Random Structures and Algorithms* 6:161 (1995).

Molloy, M. and Reed, B. The size of the giant component of a random graph with a given degree sequence. *Combinatorics, Probability and Computing* 7:295 (1998).

Montoya, J. and Solé, R. Topological properties of food webs: From real data to community assembly models. *Oikos* 102:614 (2003).

Mukhtar, M. S., Carvunis, A.-R., Dreze, M., et al. Independently evolved virulence effectors converge onto hubs in a plant immune system network. *Science* 29:596 (2011).

Murphy, K. and Mian, S. Modelling Gene Expression Data Using Dynamic Bayesian Networks. Tech. report, University of California (1999).

Navlakha, S. and Kingsford, C. Network archaeology: Uncovering ancient networks from present-day interactions. *PLoS Computational*

Biology 7:e1001119 (2011).

Nechaev, S. and Severinov, K. Inhibition of *Escherichia coli* RNA polymerase by bacteriophage T7 gene 2 protein. *Journal of Molecular Biology* 289:815 (1998).

Neidhardt, F. C., Curtiss, R., Ingraham, J. L., et al., Eds. *Escherichia coli and Salmonella: Cellular and Molecular Biology.* American Society for Microbiology (1996).

Newman, M., Strogatz, S. H., and Watts, D. J. Random graphs with arbitrary degree distributions and their applications. *Physical Review E* 64:026118 (2001).

Newman, M. E. J. Analysis of weighted networks. *Physical Review E* 70:056131 (2004).

Newman, M. E. J. Modularity and community structure in networks. *Proceedings of the National Academy of Sciences USA* 103:8577 (2006).

Newman, M. E. J. Random graphs with clustering. *Physical Review Letters* 103:058701 (2009).

Newman, M. E. J. and Girvan, M. Finding and evaluating community structure in networks. *Physical Review E* 69:026113 (2004).

Nitsch, D., Goncalves, J. P., Ojeda, F., de Moor, B., and Moreau, Y. Candidate gene prioritization by network analysis of differential expression using machine learning approaches. *BMC Bioinformatics* 11:460 (2010).

Nitsch, D., Tranchevent, L.-C., Thienpont, B., et al. Network analysis of differential expression for the identification of disease-causing genes. *PLoS ONE* 4:e5526 (2009).

Oppenheim, A. B., Kobiler, O., Stavans, J., Court, D. L., and Adhya, S. Switches in bacteriophage *lambda* development. *Annual Review of Genetics* 39:409 (2005).

Orth, J. D., Thiele, I., and Palsson, B. O. What is flux balance analysis? *Nature Biotechnology* 28:245 (2010).

Oti, M., Snel, B., Huynen, M. A., and Brunner, H. G. Predicting disease genes using protein-protein interactions. *Journal of Medical Genetics* 43:691 (2006).

Paladugu, S. R., Zhao, S., Ray, A., and Raval, A. Mining protein networks for synthetic genetic interactions. *BMC Bioinformatics* 9:426 (2008).

Palsson, B. O. *Systems Biology: Properties of Reconstructed Networks.* Cambridge University Press, Cambridge, MA (2006).

Papp, B., Pal, C., and Hurst, L. D. Dosage sensitivity and the evolution of gene families in yeast. *Nature* 424:194 (2003).

Park, C. Y., Hess, D. C., Huttenhower, C., and Troyanskaya, O. G. Simultaneous genome-wide inference of physical, genetic, regulatory, and functional pathway components. *PLoS Computational Biology* 6:e1001009 (2010).

Park, J., Lee, D.-S., Christakis, N. A., and Barabási, A.-L. The impact of cellular networks on disease comorbidity. *Molecular Systems Biology* 5:262 (2009).

Park, S., Yang, J.-S., Shin, Y.-E., et al. Protein localization as a principal feature of the etiology and comorbidity of genetic diseases. *Molecular Systems Biology* 7:494 (2011).

Pazos, F. and Valencia, A. *In silico* two-hybrid system for the selection of physically interacting protein pairs. *Proteins* 47:219 (2002).

Perrin, B.-E., Ralaivola, L., Mazurie, A., et al. Gene networks inference using dynamic Bayesian networks. *Bioinformatics* 19, Suppl. 2:ii138 (2003).

Petti, A. and Church, G. M. A network of transcriptionally coordinated functional modules in *Saccharomyces cerevisiae. Genome Research* 15:1298 (2005).

Polya, G. Probabilities in proofreading. *The American Mathematical Monthly* 83:42 (1976).

Portis, T., Grossman, W. J., Harding, J. C., Hess, J. L., and Ratner, L. Analysis of *p53* inactivation in a human T-cell leukemia virus type 1 Tax transgenic mouse model. *Journal of Virology* 75:2185 (2001).

Przulj, N. and Higham, D. Modeling protein-protein interaction networks via a stickiness index. *Journal of the Royal Society Interface* 3:711 (2006).

Ptashne, M. Regulation of transcription: From *lambda* to eukaryotes.

Trends in Biochemical Sciences 30:275 (2005).

Qi, Y., Balem, F., Faloutsos, C., Klein-Seetharaman, J., and Bar-Joseph, Z. Protein complex identification by supervised graph local clustering. *Bioinformatics* 24:i250 (2008).

Qi, Y., Bar-Joseph, Z., and Klein-Seetharaman, J. Evaluation of different biological data and computational classification methods for use in protein interaction prediction. *Proteins* 63:490 (2006).

Rambaldi, D., Georgi, F. M., Capuani, F., Ciliberto, A., and Ciccarelli, F. D. Low duplicability and network fragility of cancer genes. *Trends in Genetics* 24:427 (2008).

Rangel, C., Angus, J., Ghahramani, Z., et al. Modeling T-cell activation using gene expression profiling and state-space models. *Bioinformatics* 20:1361 (2004).

Raval, A. Some asymptotic properties of duplication graphs. *Physical Review E* 68:066119 (2003).

Reguly, T., Breitkreutz, A., Boucher, L., and Breitkreutz, B.-J. Comprehensive curation and analysis of global interaction networks in *S. cerevisiae*. *Journal of Biology* 5:11 (2006).

Riley, R., Lee, C., Sabatti, C., and Eisenberg, D. Inferring protein domain interactions from databases of interacting proteins. *Genome Biology* 6:R89 (2005).

Roach, J. C., Glusman, G., Smit, A. F. A., et al. Analysis of genetic inheritance in a family quartet by whole-genome sequencing. *Science* 328:636 (2010).

Sawhill, B. K. Genetic Function Analysis. *Santa Fe Institute Working Paper* (1995).

Scholl, C., Fröhling, S., Dunn, I. F., et al. Synthetic lethal interaction between oncogenic *KRAS* dependency and *STK33* suppression in human cancer cells. *Cell* 137:821 (2009).

Schuldiner, M., Collins, S. R., Thompson, N. J., et al. Exploration of the function and organization of the yeast early secretory pathway through an epistatic miniarray profile. *Cell* 123:507 (2005).

Segré, D., DeLuna, A., Church, G. M., and Kishony, R. Modular epistasis in yeast metabolism. *Nature Genetics* 37:77 (2005).

Segré, D., Vitkup, D., and Church, G. M. Analysis of optimality in natural and perturbed metabolic networks. *Proceedings of the National Academy of Sciences USA* 99:15112 (2002).

Shannon, C. E. and Weaver, W. *The Mathematical Theory of Communication*. University of Illinois Press, Champaign (1963).

Sharan, R., Ideker, T., Kelley, B., Shamir, R., and Karp, R. M. Identification of protein complexes by comparative analysis of yeast and bacterial protein interaction data. *Journal of Computational Biology* 12:835 (2005).

Shou, C., Bhardwaj, N., Lam, H. Y. K., et al. Measuring the evolutionary rewiring of biological networks. *PLoS Computational Biology* 7:e1001050 (2011).

Skrabanek, L., Saini, H. K., Bader, G. D., and Enright, A. J. Computational prediction of protein-protein interactions. *Molecular Biotechnology* 38:1 (2008).

Smits, W. K., Kuipers, O. P., and Veening, J.-W. Phenotypic variation in bacteria: The role of feedback regulation. *Nature Reviews Microbiology* 4:259 (2006).

Snitkin, E. S. and Segré, D. Epistatic interaction maps relative to multiple metabolic phenotypes. *PLoS Genetics* 7:e1001294 (2011).

Solé, R. and Pastor-Satorras, R. Complex networks in genomics and proteomics. In S. Bornholdt and H. G. Schuster, Eds., *Handbook of Graphs and Networks: From the Genome to the Internet*. Wiley-VCH, Berlin (2003).

Stark, C., Breitkreutz, B. J., Reguly, T., et al. BIOGRID: A general repository for interaction datasets. *Nucleic Acids Research* 34:D535 (2012).

Stelzl, U., Worm, U., Lalowski, M., et al. A human protein-protein interaction network: A resource for annotating the proteome. *Cell* 122:957 (2005).

Taylor, I. W., Linding, R., Warde-Farley, D., et al. Dynamic modularity in protein interaction networks predicts breast cancer outcome. *Nature Biotechnology* 27:199 (2009).

Taylor, J., Braasch, I., Frickey, T., Meyer, A., and van de Peer, Y.

Genome duplication, a trait shared by 22,000 species of ray-finned fish. *Genome Research* 13:382 (2003).

Taylor, J. and Raes, J. Duplication and divergence: The evolution of new genes and old ideas. *Annual Reviews in Genetics* 38:615 (2004).

Tong, A. H. Y., Lesage, G., Bader, G. D., et al. Global mapping of the yeast genetic network. *Science* 303:808 (2004).

Uetz, P., Giot, L., Cagney, G., et al. A comprehensive analysis of protein-protein interactions in *Saccharomyces cerevisiae*. *Nature* 403:623 (2000).

van Dongen, S. *Graph Clustering by Flow Simulation*. Ph.D. thesis, University of Utrecht (2000).

van Driel, M. A., Bruggeman, J., Vriend, G., Brunner, H. G., and Leunissen, J. A. M. A text-mining analysis of the human phenome. *European Journal of Human Genetics* 14:535 (2006).

van Nimwegen, E. Scaling laws in the functional content of genomes. *Trends in Genetics* 9:479 (2003).

Varma, A. and Palsson, B. O. Stoichiometric flux balance models quantitatively predict growth and metabolic by-product secretion in wild-type *Escherichia coli W3110*. *Applied and Environmental Microbiology* 60:3724 (1994).

Vázquez, A., Flammini, A., Maritan, A., and Vespignani, A. Modeling of protein interaction networks. *ComPlexUs* 1:38 (2003).

Vision, T., Brown, D., and Tanksley, S. The origins of genomic duplication in *Arabidopsis*. *Science* 290:2114 (2000).

von Dassow, G., Meir, E., Munro, E. M., and Odell, G. M. The segment polarity network is a robust developmental module. *Nature* 406:188 (2000).

Wagner, A. The yeast protein interaction network evolves rapidly and contains few redundant duplicate genes. *Molecular Biology and Evolution* 18:1283 (2001).

Wagner, A. How the global structure of protein interaction networks evolves. *Proceedings of the Royal Society of London B* 270:457 (2003).

Walhout, A. J., Sordella, R., Lu, X., et al. Protein interaction mapping in *C. elegans* using proteins involved in vulval development. *Science*

287:116 (2000).

Wang, J., Zhang, S., Wang, Y., Chen, L., and Zhang, X.-S. Disease-aging network reveals significant roles of aging genes in connecting genetic diseases. *PLoS Computational Biology* 5:e1000521 (2009).

Wang, W., Yu, H., and Long, M. Duplication-degeneration as a mechanism of gene fission and the origin of new genes in *Drosophila* species. *Nature Genetics* 36:523 (2004).

Wass, M. N., Fuentes, G., Pons, C., Pazos, F., and Valencia, A. Towards the prediction of protein interaction partners using physical docking. *Molecular Systems Biology* 7:469 (2011).

Watts, D. and Strogatz, S. Collective dynamics of 'small-world' networks. *Nature* 393:440 (1998).

Wilks, S. *Mathematical Statistics*. Wiley, New York (1962).

Wolfe, K. and Shields, D. Molecular evidence for an ancient duplication of the entire yeast genome. *Nature* 387:708 (1997).

Wong, S. L., Zhang, L. V., Tong, A. H. Y., et al. Combining biological networks to predict genetic interactions. *Proceedings of the National Academy of Sciences* 101:15682 (2004).

Wu, G., Feng, X., and Stein, L. A human functional protein interaction network and its application to cancer data analysis. *Genome Biology* 11:R53 (2010).

Wu, X., Jiang, R., Zhang, M. Q., and Li, S. Network-based global inference of human disease genes. *Molecular Systems Biology* 4:189 (2008).

Xia, J., Sun, J., Jia, P., and Zhao, Z. Do cancer proteins really interact strongly in the human protein-protein interaction network? *Computational Biology and Chemistry* 35:121 (2011).

Yu, H., Luscombe, N. M., Lu, H. X., et al. Annotation transfer between genomes: Protein-protein interologs and protein-DNA regulogs. *Genome Research* 14:1107 (2004).

Zhu, X.-M., Yin, L., Hood, L., and Ao, P. Robustness, stability and efficiency of phage λ genetic switch: Dynamical structure analysis. *Journal of Bioinformatics and Computational Biology* 2:785 (2004).

Zou, M. and Conzen, S. D. A new dynamic Bayesian network (DBN) approach for identifying gene regulatory networks from time course microarray data. *Bioinformatics* 21:71 (2005).

Index